T0074868

Artificial Intelligence in Cyber-Physical Systems

Artificial Intelligence (AI) and the Internet of Things (IoT) are growing rapidly in today's business world. In today's era, 25 billion devices, including machines, sensors, and cameras, are connected and continue to grow steadily. It is assumed that in 2025, 41.6 billion IoT devices will be connected, generating around 79.4 zettabytes of data.

IoT and AI are intersecting in various scenarios. IoT-enabled devices are generating a huge amount of data, and with the help of AI, this data is used to build various intelligent models. These intelligent models are helpful in our daily lives and make the world smarter.

Artificial Intelligence in Cyber-Physical Systems: Principles and Applications addresses issues related to system safety, security, reliability, and deployment strategies in healthcare, military, transportation, energy, infrastructure, smart homes, and smart cities.

Wireless Communications and Networking Technologies: Classifications, Advancement and Applications

Series Editor: D.K. Lobiyal, R.S. Rao, and Vishal Jain

The series addresses different algorithms, architecture, standards and protocols, tools and methodologies which could be beneficial in implementing next generation mobile network for the communication. Aimed at senior undergraduate students, graduate students, academic researchers and professionals, the proposed series will focus on the fundamentals and advances of wireless communication and networking, and their such as mobile ad-hoc network (MANET), wireless sensor network (WSN), wireless mess network (WMN), vehicular ad-hoc networks (VANET), vehicular cloud network (VCN), vehicular sensor network (VSN) reliable cooperative network (RCN), mobile opportunistic network (MON), delay tolerant networks (DTN), flying ad-hoc network (FANET) and wireless body sensor network (WBSN).

Cloud Computing Enabled Big-Data Analytics in Wireless Ad-hoc Networks
Sanjoy Das, Ram Shringar Rao, Indrani Das, Vishal Jain, and Nanhay Singh

Smart Cities: Concepts, Practices, and Applications
Krishna Kumar, Gaurav Saini, Duc Manh Nguyen, Narendra Kumar, and Rachna Shah

Wireless Communication: Advancements and Challenges
Prashant Ranjan, Ram Shringar Rao, Krishna Kumar, and Pankaj Sharma

Wireless Communication with Artificial Intelligence: Emerging Trends and Applications
Anuj Singal, Sandeep Kumar, Sajjan Singh, and Ashish Kr. Luhach

Computational Intelligent Security in Wireless Communications
Suhel Ahmad Khan, Rajeev Kumar, Omprakash Kaiwartya, Raees Ahmad Khan, and Mohammad Faisal

Networking Technologies in Smart Healthcare: Innovations and Analytical Approaches
Pooja Singh, Omprakash Kaiwartya, Nidhi Sindhwani, Vishal Jain, and Rohit Anand

Artificial Intelligence in Cyber Physical Systems: Principles and Applications
Anil Kumar Sagar, Parma Nand, Neetesh Kumar, Sanjoy Das, and Subrata Sahana

For more information about this series, please visit: https://www.routledge.com/Wireless%20Communications%20and%20Networking%20Technologies/book-series/WCANT

Artificial Intelligence in Cyber-Physical Systems

Principles and Applications

Edited by
Anil Kumar Sagar
Parma Nand
Neetesh Kumar
Sanjoy Das
Subrata Sahana

CRC Press
Taylor & Francis Group
Boca Raton London New York

CRC Press is an imprint of the
Taylor & Francis Group, an **informa** business

First edition published 2023
by CRC Press
6000 Broken Sound Parkway NW, Suite 300, Boca Raton, FL 33487-2742

and by CRC Press
4 Park Square, Milton Park, Abingdon, Oxon, OX14 4RN

CRC Press is an imprint of Taylor & Francis Group, LLC

ISBN: 9781032164830 (hbk)
ISBN: 9781032164847 (pbk)
ISBN: 9781003248750 (ebk)

DOI: 10.1201/9781003248750

Typeset in Times
by codeMantra

Contents

List of Figures

List of Tables

Preface

The cyber-physical systems (CPS) is an emerging computing concept that opened many research domains for academicians and industries worldwide. In our present living and future prospect everywhere, dependency on Artificial Intelligence (AI) and Internet of Things (IoT) increases tremendously. In each aspect, our applications are becoming smart and intelligent based. Applications of IoT include smart cities, agriculture, healthcare, home automation, water supply management, intelligent vehicles, etc. With the help of wireless communications, IoT is easily manageable for all of us.

In IoT, we can consider AI an integral bundle of several arranged objects. From this point of view, it is fundamental to comprehend the part of AI which will give a worldwide spine to the overall data sharing. Hence, it is likewise basic to consider new empowering advances, for example, remote sensor organizations, intelligent 5G technology, and computer vision. This book aims to introduce CPS with wide application areas including AI procedures and other rising canny methodologies for worldwide enhancement and enrichment of IoT. The utilization of AI in IoT, assessment measurements, limitations, and open issues about the tended-to themes are included. This book will help specialists and experts working in the domain of AI. This book also covers various issues related to CPS and how higher system performance can be achieved by improving wireless device sensing capability, real-time data processing, and predicting system behaviour.

Editors

Dr. Anil Kumar Sagar is currently working as a professor in the Department of Computer Science and Engineering, Sharda University Greater Noida, India. He did his B.E., M.Tech., and Ph.D. in Computer Science. Before joining Sharda University, he has worked as a professor at the School of Computing Science and Engineering, Galgotias University, India. His current research interests include Mobile Ad Hoc Networks and Vehicular Ad Hoc Networks, IoT, and AI. He has published numerous papers in international journals and conferences including IEEE and Springer. He has received a Young Scientist Award for the year 2018–19 from the Computer Society of India and the Best Faculty Award for the years 2006 and 2007 from SGI, Agra.

Dr. Parma Nand has received his Ph.D. in Computer Science & Engineering from IIT Roorkee, and M.Tech. and B.Tech. in Computer Science & Engineering from IIT Delhi. Prof. Parma Nand is having more than 27 years of experience both in industry and in academia. He has received various awards like the Best Teacher Award from the Union Minister, Best Students' Project Guide Award from Microsoft in 2015, and Best Faculty Award from Cognizant in 2016. He had successfully completed government-funded projects and spearheaded the last five IEEE International Conferences on Computing, Communication & Automation, IEEE Students Chapters, Technovation Hackathon 2019, Technovation Hackathon 2020, and International Conference on Computing, Communication, and Intelligent Systems (ICCCIS-2021). He is a member of Executive Council of IEEE UP section (R-10), member Executive Committee IEEE Computer and Signal Processing Society, member of Exec. India Council Computer Society, and member of Executive Council Computer Society of India, Noida section and has acted as an observer in many IEEE conferences. He is also having active memberships of ACM, IEEE, CSI, ACEEE, ISOC, IAENG, and IASCIT. He is a life-time member of Soft Computing Research Society and ISTE.

Dr. Sanjoy Das is working as a professor and head of the Department of Computer Science, Indira Gandhi National Tribal University (A Central Government University), Amarkantak, M.P. (Regional Campus Manipur), India. He did his B.E., M.Tech., and Ph.D. in Computer Science. He did his Ph.D. from Jawaharlal Nehru University, New Delhi. Before joining IGNTU, he has worked as an associate professor at the School of Computing Science and Engineering, Galgotias University, India. He also worked as an assistant professor at G. B. Pant Engineering College (A Govt. Institute), Uttarakhand and at Assam University (Central University), Silchar. He has 16+ years of experience in teaching and research. He has organized many international conferences and attended as session chair. He has published 70+ research papers in SCI, Scopus indexed, referred international journals and conferences, including publishers IEEE, Inderscience, Elsevier, and Springer. He has been the editor of five books on cloud computing, vehicular ad hoc networks, etc. His current research interests include mobile ad hoc networks and vehicular ad hoc networks, distributed systems, and data mining.

Dr. Neetesh Kumar received his M.Tech. and Ph.D. degrees from the School of Computer and Systems Sciences, Jawaharlal Nehru University, New Delhi. He has qualified GATE, and NET+JRF. He has got Travel Grant Awarded from IEEE IPDPS International Society and Special Mention Awarded from Xerox Research Centre, Bangalore. He has worked as an assistant professor in Shri Mata Vaishno Devi University (SMVDU), Katra, Jammu & Kashmir, India. He has also been appointed as a faculty incharge of Computer Network Center at SMVDU. Later, he has worked as an assistant professor in Delhi Technological University, Delhi. He has also been a faculty member at the Atal Bihari Vajpayee-Indian Institute of Information Technology and Management (IIIT) Gwalior, India. Currently, he is working as a faculty member at IIT Roorkee, India. He has published many research publications in world's top tier publishers like IEEE journals and transactions, Elsevier journals (including *Future Generation Computer System* (FGCS), *Information Sciences*, etc.) and Springer journals (*Soft Computing, Computing*, etc.). One of his articles published in IEEE's *Sensors* journal has been notified by IEEE Council in the list of world's top 15 most downloaded articles in the month of October–November 2018. He is a regular reviewer of highly reputed journals like *IEEE Transactions on Parallel and Distributed Systems, IEEE Journal of Internet of Things, FGCS*, etc. He is also acting as a lead PI for three sponsored projects from DST/CSIR agencies, Government of India. He has been a technical program committee member in several conferences. He has also been invited as a keynote speaker in the conference held at Amity Gwalior and conference chair at IEEE CICT-2019 at IIIT Allahabad. Broadly, his research interests include algorithm design, scheduling in parallel computing systems (multi-core/many-core systems), applied evolutionary computing, and intelligent transportation systems.

Dr. Subrata Sahana is an associate professor in the Department of CSE at Sharda University, Greater Noida, U.P. He received his Ph.D. in Computer Science from the JNU, New Delhi (Central University), India (NIRF Ranking-2020: 02), M.Tech. in Computer Science and Engineering form BIT, Mesra in 2010 (NIRF Ranking-2020 under 100), and B.Tech. in Computer Science & Engineering from CEMK, Kolaghat affiliated from West Bengal University of Technology in 2007. He joined Sharda University as an assistant professor in August 2019 and promoted as an associate professor in January 2021. Prior to this, he was associated with Galgotias University as an assistant professor from 2012 to 2019, V.I.T. University, Vellore as an assistant professor from 2010 to 2012 and RVSCET, Jamshedpur as a Lecturer from 2007 to 2008. He has rich experience in publication in many international journals and conferences with high repute. He is also serving many reputed journals as editorial board member and reviewer board member. Moreover, Dr. Sahana has also delivered expert talks and guest lectures in international conferences and serving as the reviewer for journals of IEEE, Springer, IGI Global, etc. His research areas are underwater wireless sensor networks, pattern matching, bio-informatics, IoT, security, and cryptography.

Contributors

Abdul Basit Aftab
Mechatronics Engineering Department
Shaheed Zulfikar Ali Bhutto Institute of
    Science and Technology (SZABIST)
Karachi, Pakistan

Imran Amin
Mechatronics Engineering Department
Shaheed Zulfikar Ali Bhutto Institute of
    Science and Technology (SZABIST)
Karachi, Pakistan

Darpan Anand
Chandigarh University
Mohali, India

Pratul Arvind
Dr. Akhilesh Das Gupta Institute of
    Technology and Management
New Delhi, India

Mallavarapu Rajan Babu
Lendi Institute of Engineering and
    Technology
Vizianagaram, India

Kiranmai Babburu
Lendi Institute of Engineering and
    Technology
Vizianagaram, India

Ajay Kumar Balmiki
Maulana Abul Kalam Azad University
    of Technology
West Bengal, India

A M Anusha Bamini
Department of Computer Science
    Engineering
Karunya Institute of Technology and
    Sciences
Coimbatore, Tamil Nadu

Parul Bhanarkar
Jhulelal Institute of Technology
Nagpur, India

Tushar Bhardwaj
Applied Research Center
Florida International University
Miami, Florida

Parijat Bhowmick
Indian Institute of Technology
Guwahati, India

Pradnya S Borkar
Jhulelal Institute of Technology
Nagpur, India

K Guru Charan
Lendi Institute of Engineering and
    Technology
Vizianagaram, India

Sanjoy Das
Indira Gandhi National Tribal
    University Regional Campus
    Manipur
Imphal, India

Sima Das
Maulana Abul Kalam Azad University
    of Technology
West Bengal, India

Divya Gangwani
Department of Electric Engineering
    and Computer Science
Florida Atlantic University
Boca Raton, Florida

Pranav Gangwani
Department of Electrical
    and Computer
    Engineering
Florida International University
Miami, Florida

Pratiksha Gautam
Amity University
Gwalior, India

Akash Goel
Raj Kumar Goel Institute of
    Technology
Ghaziabad, India

Nidhi Gupta
Raj Kumar Goel Institute of
    Technology
Ghaziabad, India

Gurucharan Kapila
Lendi Institute of Engineering and
    Technology
Vizianagaram, India

Shailesh Kumar Gupta
Raj Kumar Goel Institute of
    Technology
Ghaziabad, India

J H Jensha Haennah
St. Xavier's Catholic College of
    Engineering
Anna University
Chennai, India

Ashu Jain
Dr. Akhilesh Das Gupta Institute of
    Technology and Management
New Delhi, India

Dhyanendra Jain
Dr. Akhilesh Das Gupta Institute of
    Technology and Management
New Delhi, India

Faraz Junejo
Mechatronics Engineering Department
Shaheed Zulfikar Ali Bhutto Institute of
    Science and Technology (SZABIST)
Karachi, Pakistan

Amit Kumar Pandey
Dr. Akhilesh Das Gupta Institute of
    Technology and Management
New Delhi, India

Jaishanker Prasad Keshari
IIMT College of Engineering
Greater Noida, India

Imran Ahmed Khan
Jamia Millia Islamia
New Delhi, India

Amruta Khot
Walchand College of Engineering
Sangli, India

G R Gnana King
Sahrdaya College of Engineering and
    Technology
Kodakara, India

S S Kiran
Lendi Institute of Engineering and
    Technology
Vizianagaram, India

B Kiranmai
Lendi Institute of Engineering and
    Technology
Vizianagaram, India

Manoj Kumar
Amity University
Gwalior, India

Neetesh Kumar
IIT Roorkee
Roorkee, India

Sandeep Kumar
IIMT College of Engineering
Greater Noida, India

Leonel Lagos
Applied Research Center
Florida International University
Miami, Florida

Akhilesh Latoria
AURO University
Surat, India

Kaushik Mazumdar
Maulana Abul Kalam Azad
 University of Technology
West Bengal, India

Nidhi
Mata Sundri College for Women
University of Delhi
New Delhi, India

Prashant Panse
Medi-Caps University
Indore, India

Alexander Perez-Pons
Department of Electrical and Computer
 Engineering
Florida International University
Miami, Florida

Anushka Purwar
Mata Sundri College for Women
University of Delhi
New Delhi, India

Ranjit Rajak
Dr. Harisingh Gour Central
 University
Sagar, India

Naveen Rathee
IIMT College of Engineering
Greater Noida, India

Varnika Rathee
VIT
Bhopal, India

Atif Saeed
Mechatronics Engineering
 Department
Shaheed Zulfikar Ali Bhutto Institute of
 Science and Technology (SZABIST)
Karachi, Pakistan

Anil Kumar Sagar
Sharda University
Greater Noida, India

Mohammad Sajid
Aligarh Muslim University
Aligarh, India

Vijay Bhaskar Semwal
NIT
Bhopal, India

Kaljot Sharma
Chandigarh University
Mohali, India

Jagendra Singh
Bennett University
Greater Noida, India

Monika Singh
Brahmanand Mahavidhyalaya
Bulandshahr, India

Neha Singh
Raj Kumar Goel Institute of
 Technology
Ghaziabad, India

Prashant Singh
Dr. Akhilesh Das Gupta Institute of
 Technology and Management
New Delhi, India

Gaurav Sinha
Department of Electronics and
    Communication Engineering
IIMT College of Engineering,
    Knowledge Park-III
Greater Noida, India

Dinesh Soni
IIT Roorkee
Roorkee, India

Reena Thakur
Jhulelal Institute of Technology
Nagpur, India

Himanshu Upadhyay
Applied Research Center
Florida International University
Miami, Florida

About This Book

Chapter 1: Mood-Detection Based Media Recommendation

Prashant Singh, Dhyanendra Jain, Ashu Jain, Pratul Arvind, Amit Kumar Pandey, Anil Kumar Sagar, and Sanjoy Das

This chapter proposes a way to use different technologies to detect mood and recommend a media source, like images, music, videos, podcasts, poems, etc., suiting to their current mood. Moodiest is one such Android-based application that use machine-learning–based NLP and CNN algorithms to detect emotion and recommend media accordingly.

Chapter 2: AI- and IoT-Based Body Sensor Networks for Healthcare System: A Systematic Review

Divya Gangwani

This chapter aims to provide a review of an intelligent healthcare system based on IoT and deep learning technologies in AI, which can identify serious health issues in patients and provide an option of low-cost treatment and hospital administration.

Chapter 3: On the Convergence of Blockchain and IoT for Enhanced Security

Pranav Gangwani, Tushar Bhardwaj, Alexander Perez-Pons, Himanshu Upadhyay, and Leonel Lagos

In this book chapter, we are going to talk about the various security issues with IoT and how the blockchain integration can address those security risks. Additionally, we will also present the challenges that need to be addressed by integrating IoT with blockchain technology.

Chapter 4: Artificial Intelligence in Cloud Computing

Dinesh Soni and Neetesh Kumar

This chapter presents relevant literature describing practical applications, structure, components, characteristics, programing languages, advantages, domains, challenges, security, intrusion detection, smart grid, and dominating research topics in CPS. In addition, the mapping of CPS for the healthcare domain that will help in trends visualization, event prediction in healthcare, and determining the differences between cloud-based healthcare solutions with CPS-based healthcare solutions are also presented.

Chapter 5: Current and Future Trends in Intelligent Transportation System with Applications of AI

Atif Saeed, Abdul Basit Aftab, Faraz Junejo, and Imran Amin

This chapter covers a collection of challenges affecting the transportation industry that is categorized as intelligent transportation systems. Traffic management, public transportation, safety management, manufacturing, and logistics are some of the subsystems considered in intelligent transportation systems where AI benefits are used.

Chapter 6: Intelligent 5G Networks and Augmented Virtual Reality in Smart Transportation

A M Anusha Bamini, G R Gnana King, and J H Jensha Haennah

This work provides knowledge of communication technologies and paradigms for connected and automated mobility for smart roads and vehicles. It shows how and where to use these technologies to provide innovative services on the road that would otherwise be impossible. At the same time, a special focus will be given to the mobility services that can be offered using 5G mobile technologies, such as in-vehicle augmented virtual reality and road safety services.

Chapter 7: Cyber-Physical Security Issues and Challenges Using Machine Learning and Deep Learning Technologies

Sima Das, Ajay Kumar Balmiki, Kaushik Mazumdar, and Parijat Bhowmick

The survey presented in this chapter will also give the idea of certain valuable inducements regarding security or security threat. This chapter describes how the challenges have arrived in the IoT world and discuss possible solutions to overcome the deep-down of the stranded.

Chapter 8: Brain MRI Image Active Contour Segmentation for Healthcare Systems

Kiranmai Babburu, Gurucharan Kapila, S S Kiran, and Mallavarapu Rajan Babu

In this chapter, an active contour Chan-Vese segmentation algorithm has been used. This algorithm comprises two energies, of which one power diminishes the form and another expansion of the shape. These two energies get adjusted when the form arrives at the furthest point of our ideal item, which brings about total segmentation.

Chapter 9: Machine Learning Techniques Applied to Extract Objects from Images: Research Issues Challenges and a Case Study

Reena Thakur, Pradnya S Borkar, Parul Bhanarkar, and Prashant Panse

This chapter describes various machine learning techniques applied to extract objects from images and case studies. First, the related literature works followed by background knowledge are presented. In addition, how machine learning (ML) algorithms are applied in the realm of extracting objects from images, from the perception of optimization techniques, resource management, and security has been reviewed. Finally, case studies are discussed, which are worthwhile to be pursued by the researchers in the future.

Chapter 10: AI and IoT-Enabled Technologies and Applications for Smart City

Shailesh Kumar Gupta and Neha Singh

This chapter describes how AI and IoT can be used to make cities smarter. The chapter begins with a discussion of AI and the IoT fundamentals using a hybrid approach. Following that, AI- and IoT-enabled applications and technologies are discussed. How IoT and AI may be viewed as enabling technologies for smart cities is also discussed. Developing a smart city using AI and IoT techniques requires many issues to be appropriately addressed.

Chapter 11: Blood Cancer Classification with Gene Expression Using Modified Convolutional Neural Network Approach

Nidhi Gupta, Akhilesh Latoria, and Akash Goel

The work presented in this chapter aims to design a supervised machine learning approach to differentiate between AML and ALL, i.e., acute myeloid leukemia and acute lymphoblastic leukemia, using modified convolutional neural network approach.

Chapter 12: An Introspective Approach to Fathom Human-Inspired Bipedal Walk Using Gait Analysis for the Matrix of Cyber-Physical System

Manoj Kumar, Pratiksha Gautam, and Vijay Bhaskar Semwal

This chapter is an attempt to reconcile the work done in humanoid and achieving humanoid walking in the context of a CPS. This work focuses on the fundamentals of gait analysis into robotics applications, and later CPS aspect of humanoids has been discussed with current limitations, research challenges, and future scope.

Chapter 13: Capacitated Vehicle Routing Problem Using Algebraic Particle Swarm Optimization with Simulated Annealing Algorithm

Mohammad Sajid, Jagendra Singh, and Ranjit Rajak

This work proposes an algebraic particle swarm optimization with a simulated annealing (SA) algorithm to address the CVRP to make the total traveled distance minimum. The proposed algorithm consisting of an algebraic version of PSO and SA works on the permutations and employs an algebraic composition operator to realize the different operations of PSO.

Chapter 14: An Innovative Smart IoT Device to Measure and Monitor Patient's Critical Parameters in Hospitals

Naveen Rathee, Jaishanker Prasad Keshari, Sandeep Kumar, Varnika Rathee, Gaurav Sinha, Imran Ahmed Khan, and Monika Singh

In this chapter, actual and measured data of critical parameters like patients' heart rate, temperature, pulse rate, and blood pressure have been recorded using ICU instruments in a hospital by a designed smart IoT-proposed device for measuring critical parameters of the patient in healthcare.

Chapter 15: Health Analysis by Digital Doctor Using Deep Neural Network

Amruta Khot

In this chapter, we built such a tool using the power of natural language processing and recommender system, making medical processes easy. It can not only understand doctor-patient conversations but can also work on any other datasets.

Chapter 16: AI and IoT in Supply Chain Management and Disaster Management

Kaljot Sharma and Darpan Anand

This chapter describes the role of AI and IoT in supply chain management agreements and disaster management. Also, it discusses about the importance and benefits of these technologies along with the challenges of using these technologies? How is it essential to use them in supply chain management and disaster agreement? The

AI platform improves results by evaluating and analyzing all available data and best practices in order to give insights into and help in decision-making.

Chapter 17: Cyborgs: A Coming Era

Anushka Purwar and Nidhi

This chapter discusses how cyborgs arose with the help of AI and cybernetics, applications of cybernetics, and different types of cyborgs present. The key contribution of this study is the presentation of cyborgization, the increase of biological beings with capabilities, and its future aspects as the effects of increased advancement in technology.

1 Mood-Detection Based Media Recommendation

Prashant Singh, Dhyanendra Jain, Ashu Jain, Pratul Arvind, and Amit Kumar Pandey
Dr. Akhilesh Das Gupta Institute of
Technology and Management

Anil Kumar Sagar
Sharda University Greater Noida

Sanjoy Das
Indira Gandhi National Tribal University
Regional Campus Manipur

CONTENTS

1.1 INTRODUCTION

The temperament of a human is generally reliant upon the feelings he/she has throughout an unequivocal timeframe. A human can distinguish the feelings of another human by directly seeing his face or by conversing with him and noticing the way he talks, the words he utilizes, and the demeanors he portrays while talking. These days, this capacity is being studied by machines utilizing profound learning methods where calculations are planned so that they take input as text, pictures, or reviews and find the feelings, not to mention the disposition, of the client. Likewise, various applications are carried out alongside suggesting clients with activities and treatment

DOI: 10.1201/9781003248750-1

remedies to assist them with handling their disposition. These kinds of calculations and applications are extremely valuable in clinical scenarios as once in a while a therapist thinks that it is hard to distinguish a patient's state of mind; subsequently, he can undoubtedly depend on an ideal machine-based disposition identification framework. Ordinarily utilized calculations are carried out by an NLP (natural language processor) that handles the info given as text, incorporating various technologies like preprocessing, eliminating stop words, stemming, lemmatization, and mark encoding. We use characterization calculations like k – nearest neighbors irregular woods, Naive Bayes, and logistic regression to decide the temperament. On account of profile pictures, CNNs (convolutional neural networks) are utilized to study them, and this course of extraction in which the mindset is supposed to dependent on elements of the face is called include extraction. This calculation is a significant piece of profound learning. In this chapter, we present an application that performs calculations based on NLP and committee CNNs to recover the client temperament and utilizes an artificial intelligence (AI) method to decide the mindset effects of various media like music, recordings, and pictures and map them together to plan the final list of suggested media.

Vertical push as an element of the substantial arousing that is both non-public and individual. Eye-to-eye addresses play a significant role in sorting out an individual's sentiments and the current disposition. Machines acquiring information can find sentiments by observing the looks and examining what each statement implies. That expertise furthermore applies to the recently gained data. As of late, gadget considering has acquired force in creative and insightful guidance programs in PC. The advantages of a gadget is principally based upon the exhortation calculations and splendid advancement on the resultant three abilities such as:

- Collaborative studying ability of solid consultant
- Content basically based thoroughly sifting profound
- Transaction among abilities

Conventional guidance structures depend upon individual explicit data content, material, and records. Over the long run, we've seen an apparent blast in endeavors to utilize sentiments in explicit ways to design better quality exhortation structures. The default guidance is one of these astonishing techniques. Various media frameworks are set as default. Investigate these destinations for get-together capacities actually like people. The individual's produce directs that don't convey in expressions and deeds in loads of ways, specifically through looks. Through video exhortation programs, we can expand sentiments by computing an individual's look.

There are numerous mechanical progressions identified with advancement in music also, or we can call them the rebuilding of music information; however, we have not yet tackled this issue here. Reestablishing music information is consistently the subject of extraordinary exploration. There is a great deal of music accessible to us, which is difficult for a PC to introduce. Then again, we have additionally taken extraordinary steps in enthusiastic knowledge. Large numbers of the techniques and calculations accessible. But here as well, the issue stays annoyingly on the grounds that the two capacities are performed in an unexpected way; here it attempts to

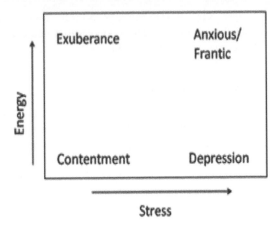

FIGURE 1.1 Thayer's Model of Mood [1].

consolidate the two innovative headways, even in their typical lives. In the standard FER framework, the eyes and mouth are significant highlights of looks. Your feelings or sentiments are reflected by your eyes and lips such as "I can't imagine you interestingly."

1.1.1 Mood Taxonomy

As mentioned in [2], the conventional approach to identifying mood reaction in music psychology is to use adjective descriptors, such as pathetic, optimistic, and gloomy. In the late 1990s, a two-dimensional mood model was introduced by Thayer (1989). This dimensional approach takes the idea that mood consists of two factors, tension (happy/anxious) and energy (calm/energetic), and splits the mood of music into four clusters: contentment, sadness, exuberance, and anxious/frantic as depicted in Figure 1.1.

Such concepts are clear and discriminatory for the four clusters, and the two-dimensional framework still offers significant computer modeling markers. It is, thus, used in our method of mood detection.

1.2 RELATED WORKS

1.2.1 Media Recommendation

In [3], Erion Ç Riccardo Coppola Eleonora Gargiulo, Marco Marengo, and Maurizio Morisio proposed a mood based on car music system which relates between the mood of the driver and mood of a song by analyzing its lyrics and hence a recommendation system.

Charulatha Vijaykumar in her proposed theory [4] explained about a system by which lyrics of songs are extracted using APIs and an algorithm is used to determine mood of the user; then, the two are matched and music is played accordingly.

In [5], Aurobind V. Iyer, Viral Pasad, Smita R. Sankhe, and Karan Prajapati proposed a system that uses the feature extraction algorithm to detect emotion from the image of a face and hence recommend a suitable piece of music.

1.2.2 Text-Based Emotion Detection

In [6], the team of Efthymios Kouloumpis, Theresa Wilson, and Johns Hopkins explored the effectiveness of linguistic features in this chapter to detect the sentiment in Twitter messages. They evaluated the utility of current lexical tools as well as features that collect knowledge about the informal and innovative vocabulary used in microblogging. They took a supervised approach to the problem, but leveraged existing Twitter data of hash tags to create training data.

The authors in [7] have predicted personality based on text posted by Twitter users, B. Y. Pratama and R. Sarnoused text classification, English and Indonesian were the languages employed. Naive Bayes, k-nearest neighbors, and support vector machine were the grouping techniques applied. Test findings revealed that Naive Bayes outperformed the other strategies marginally.

1.2.3 Image-Based Emotion Detection

In [8], from a facial image, several facial parameters were extracted and used to train several generalized and advanced neural networks. The best performing generalized and advanced neural networks were recruited into decision-making committees based on initial testing that established an optimized neural network structure for the committee. Using data derived from topics not used in preparation or initial research, the integrated committee of neural network systems was then assessed.

Kayobashi and Hara proposed and developed an active human interface [9] in which they considered six groups of official expressions, i.e., surprise, fear, disgust, anger, happiness, and sadness, thus obtaining $30 x$ and y coordinates of facial characteristic points (FCP) covering three elements of the face (eyes, eyebrows, and mouth). The facial location information was then generated and fed into the input units of a neural network; network learning was performed by a back-propagation algorithm, and a recognition test was performed.

They again proposed a system in [9] in their theory, and retrieved facial images displaying mixed facial expressions from a video tape captured by the facial image of the customer, and from those images, they obtained the facial expressions information in terms of the (x, y) coordinates of FCP. The facial image location information is then created for 19 customers and used for neural network training and recognition testing.

In [10], FaceFetch is a proposal for novel media content based on a smart framework that knows the current enthusiastic period of the individual by distinguishing looks and providing audio-visual substances to individuals. The framework can know the enthusiastic period of individuals. They offer a work area just as portable UI and audio-visual substances like music, films, and different recordings of the client's interest from the cloud. In [1], they introduced an overall system of video-related substance investigation, including video content and depictions of feelings.

The current review contains both immediate and roundabout video examination of important substances. They were extremely centered around understanding video-related substance examination. Because of this turn of events, investigation of significant video content will be vital. The motive behind the video-related substance automatically marks every video cut with its touch content. In the paper [11], they cooperate. The force of two techniques is expressed and afterward proposed: EmoWare (enthusiastic mindfulness), a customized and enthusiastic shrewd video suggestion engine. It utilizes the setting of a mindful sifting technique. Also the clients' enthusiastic responses to the suggested video are taken as a face image. What's more, an examination was done to be utilized for navigation and video corpus development with ongoing reaction dispersion.

Due to the appearance of the web, society is encountering a gigantic assortment of changes. Things are done in extremely exceptional strategies than previously. It is shocking that 90% of what is expected through 2020, roughly 1,000,000 minutes of videos will be on IP organizations in accordance to the cisco white paper agreement. Many of the obligations that have been cultivated in the past are stand-alone. This is at present done online along with shopping, checking out motion pictures or movies, or paying attention to music, should be added and after that drowning in music, choosing an occasion spot, and so forth. Thus, the field of the Internet is overwhelmed with decisions that burden clients and make intense conditions for them to make decisions. This is one of the seriously fascinating RS mediates. As referenced, RSs are programing devices or systems that help the individual with insider dynamic methods through raising freedoms. What the machine predicts the individual would do can be a direct result of virtual following. Each individual on the Internet is moving, allowing RS to look into the location of the individual when away from home. Utilize the records you get hold of to create customized ideas. Also, helping the individual within a dynamic system isn't generally the easiest RS advantage, and they assist them with developing their revenue through selling products and incrementing their employability as they gather data. Improving the per-person enjoyment is broadly utilized, but there might be no proper arrangement. They are exceptional. In this manner, there are exceptional strategies for doing that or helping the pointers. Nonetheless, each of them wants to accumulate. People use their individual identity to make a profile and choose the greatest number of choices separately for a chose individual. RS studies were initiated during the 1990s, and from that point onward, a lot of research have been completed on the topic, not only in the research organization, but also in the endeavor world. There are many online groups that rely on pointers to generate additional revenue for them. Their clients, such as Amazon.com, utilize their pointers to exploit the extended tail endeavor model. In 2006, Netflix also exhibited an enlightening information base of movies. To build a counsel calculation, experts of the subject need to initially find obscure necessities of the experts already in it. Further, the advanced category of clients gave a coin prize to the most reliable and showed the meaning of association pointers.

As an uncommon subject, a ton of studies on RS have been completed in an exploratory arrangement, and it's fundamental to conduct examines in which RS are considered inside the setting of genuine use. The machine requires individuals' records to produce real forecasts. Clearly, there are stand-alone strategies to assemble these records in a roundabout way. At the point when the essential one is utilized, the individual offers records to the machine. At the point when the essential one is utilized, the individual offers records to the machine such as deliberately accomplishing some content by emulating through a film. Neatly, the individual behavior is followed out, requiring additional individual transaction with the machine. For instance, to apply it as insights to posting sound melodies, another model, fundamental such as the introduction of the task, way to the fast improvement of innovation, is to regularly stagger on the sentiments that an individual is encountering while ingesting the substance. Programed location of sentiments is examined in a review subject known as affective computing. The name "Full of Feeling Computing" was instituted by Rosalind Picard in 1997. He characterizes it as a PC identified with, ascending from, or purposefully affecting sentiments. Research on this topic explores what are the components that affect relational correspondence. The PC, individuals and the way methods along with compassion might be utilized to offer records with roughly human impacts. Furthermore, some different investigations center within the subject states' planning and look at structures that utilize the impact on their psyche.

The improvement of emotional figuring till date has been huge, specifically in computerization. Instructions to find there are stand-alone techniques for having sentiments, similar to this, along with facial investigation, outline language, voice, biometrics, and so forth. In 2005, the exploration network accepted the meaning of sentiments in RS. Feelings are fundamental in RS on the grounds that, as demonstrated, utilizing the item is made to feel sentiments. Feelings are crucial for people; they sway every day's games and independent direction. For instance, enthusiastic acclaim impacts clients' choices of whether to apply an embraced thing. In the case of RS, because of the trouble, they have largely left out sentiments due to the stock of the arrangement or setting, quantifiable and straightforward. Until this point in time, the examinations on RS and affective computing are done independently.

Promoting cardiovascular perspectives through music is an old idea effectively utilized in an alternate setting, both in the lab establishment (i.e., Westermann, Spies, Stahl and Hesse) and this present reality establishment. The opinion player by Janssen, Van Den Broek, and Westerink, melodies are consequently recommended to work on the client's condition by a natural technique such as use of skin temperature as a chemometric measurement. First, members needed to rate their own tune assortment in valence (i.e., beauty). Next, Janssen et al. (2012) split tunes into three phases of valence: off-base, nonpartisan, and great. Then, they utilized nine tunes indiscriminately in a stage pointed toward bringing members into a condition of nonpartisanship, and reduced or raised valence. Accordingly, Janssen et al. (2012) inferred that singular tune ideas added to the skin temperature of members (chemotherapy). Over skin temperatures, they utilized skin control to portray the force of energy. The authors, van der Zwaag, Janssen, and Westerink, tracked down that the members' demeanor toward where they put us affected by (increment or diminishing strength and valence) by permitting them to encounter similarity through enthusiastic tunes. The referenced examinations utilized a compulsory condition focused on at their members, which might not have done as such. That may be the most ideal way. Suitably, individual pulse takes everybody back to a formerly depicted state. Suggestions are dependent on client inclinations, which makes a customized demeanor toward the client. Also, to portray their ideal and present status, van der Zande (2018) utilized two sorts of enthusiastic scales. Initially, a rundown of inquiries was performed based on the boost of the works of Mehrabian and Russell (1974) and Matthews, Jones, and Chamberlain. Second, it makes an enthusiastic equilibrium dependent on the interface (e.g., music). Well-known tunes by members are transferred alongside their status (i.e., power and valence fueled by Spotify). The tracks are then organized into the components (i.e., power, valence) more than seven levels, making a 7×7 network where every cell contains a track addressing an allocated state—state (first tune), wanted state (objective tune) and mindset after (finish of tune).

Feeling is characterized as a condition of response to human brain science and physiology [10]. It can create. At the point when people have various responses to various sensations, contemplations, and practices, they are exposed to interior or outside transformations. Under typical conditions, the six most normal human feelings (cheerful, furious, dismal, dread, nausea, and shock) can fundamentally change. It mirrors people's sentiments [12]. In any case, when there are some convoluted circumstances, other complex feelings will likewise happen. As quite possibly the most widely recognized outer stimulus, "Music can undoubtedly influence human feelings" [13]. What various contents mean for human feelings has consistently been an intriguing issue of exploration in the areas of brain science and music. Yet, the strategies measuring the definition and portraying feelings have consistently been one of the troubles in this kind of exploration [6]. With the improvement of innovation and the rise of AI [?]. It has shown remarkable execution in various fields [12]. As a state-of-the-art innovation, AI has shown starting accomplishment in re-enacting human correspondence, insight, and activity [8]. It incorporates superior quality teaches like bionics, brain science [9], measurements, and progressed arithmetic as the center of innovation [6], and has started to arise as in medication [9,14], schooling [7,15], social science [7], and development. Besides, in certain spaces where AI can be utilized maturely, it is astounding that the work productivity of The AI is phenomenal. The foundation of AI is the rule that reproduces the course of human reasoning and getting the hang of, making it workable for AI to recognize and get people. Feelings in various circumstances The most common way of characterizing human feelings dependent on various outside indications of individuals, for example, looks [6,16], non-verbal communication developments [8], sounds [6,9], mind waves [6,9], and skin electrical signs [14]. This is called feeling acknowledgment [15]. A fascinating review was conceived when music, feeling acknowledgment, and AI impacted one another: music feeling recognition (MER) [7]. This exploration expects to assemble a framework that can precisely and proficiently distinguish feelings that are wanted to be communicated in music through research in the field of AI. Due to the high recurrence of new music made, the quality prerequisites for music investigated by feeling in the music market are accomplished [7]. Besides, MER is a valuable part in AI examination and human feeling recreation. So MER has exceptionally high exploration esteem [7]. As the feelings in music are found by an ever-increasing number of individuals, more individuals are starting to put resources into it [6]. Nonetheless, to precisely characterize and depict the feelings in music is not difficult to pass judgment. Diverse music might have distinctive enthusiastic articulations. At various occasions, or at better places, a similar tune might show up in an unexpected way. Indeed, even an individual with weak reasoning can communicate sentiments that are felt in a piece of music. However, it may not be communicated that precisely [16]. Then, at that point, how to make machines and PCs, which don't have similar considerations as individuals, express the feelings. The capacity to replicate the music precisely has turned into a significant point in AI in the sound field [3].

1.3 PROPOSED SYSTEM

The architecture for the proposed recommendation system-based application called "Moodyst" is given below in reference with [4,5] (Figure 1.2).

In this algorithm, first the user will register/login using his username and introduce himself to the application. Next, the UI will ask for an input. The user has the choice of giving input as text or image. Depending upon the input, it will be processed by respective algorithms to detect the mood of the user. In a database, there is collection of various media, viz., music, videos, podcasts, images, stories, and news, or it redirects the user to any other interface such as YouTube or Spotify depending upon his choice. Hence, it will play the suitable media.

This algorithm basically has three major components as follows.

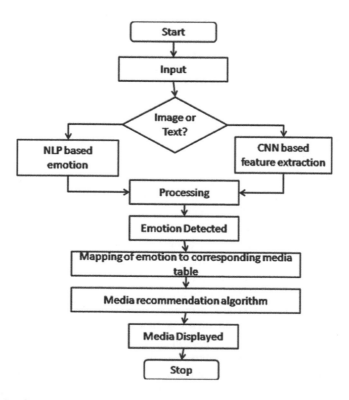

FIGURE 1.2   Overall architecture of system.

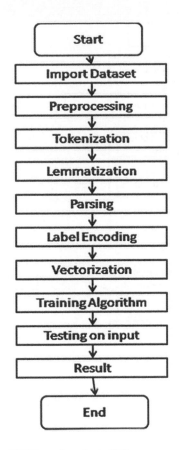

FIGURE 1.3   Flowchart on NLP based analysis [17].

1.3.1   TEXT-BASED EMOTION DETECTION

This algorithm is implemented using NLP, one of the widely used applications of machine learning (Figure 1.3).

The steps involved in this process are depicted in given flowchart with reference to [14]. The dataset used for training the algorithm is taken from Kaggle which has 30,000 rows to train and 10,000 rows to test. The dataset which is used for testing is taken from Emo Bank, 10,000 sentences annotated with valence, arousal, and dominance values. After using the first dataset to train three algorithms, viz., logistic regression, monomial Naive Bayes, and SVM, they were used upon the second dataset to find the emotion of texts in it.

The algorithm used in this part is with reference to [6,15].

1.3.2   EMOTION DETECTION USING FEATURE EXTRACTION

This algorithm is in reference to [6,8] in which we have used committee neural networks, which further use back-propagation method (Figures 1.4 and 1.5).

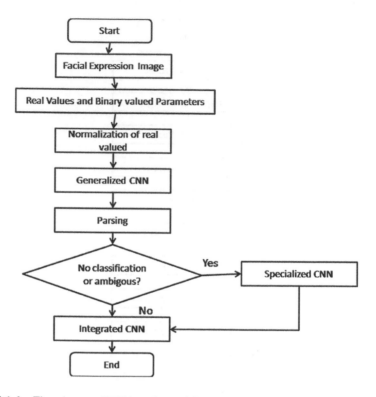

FIGURE 1.4   Flowchart on CNN based mood detection [8] and [9].

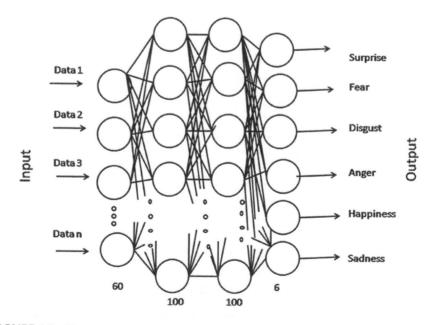

FIGURE 1.5   Neural Network NN [6].

In this algorithm, we use feature extraction [6] and committee neural networks variance of CNN [8]. We have taken a committee of CNNs, i.e., generalized and specialized CNNs. The dataset used is Cohn-Kanade database [16]. Images of facial expressions were obtained from the Cohn-Kanade database [14]. The archive featured facial photographs taken of 97 subjects aged between 18 and 30 years. Sixty-five percent of the database had female subjects. Fifteen percent of the respondents were African-American and 3% were Latino or Asian. The CNNs are first trained on the basis of real-valued and binary-valued parameters. In case of testing, image is first fed into the committee of generalized CNNs, and based on normalized value of the parameters [8], it is checked for no classification or ambiguous classification, after which it is feed to committee of specialized CNNs based on the deciding factor, and hence, the emotion detected by majority of CNNs is the final emotions.

1.3.3 MEDIA RECOMMENDATION ALGORITHM

The algorithm for recommendation is named the Moodyst algorithm. After retrieving the emotion, we will encode and label the emotion like Happy is 1001 and Sad is 1002. The media presented in the database is stored in different tables, and the names of tables match with corresponding mood table. After mapping of emotion with table, a random integer will be generated in the range of distinct values of primary key of that chosen table. On that basis, a random media will be shown, the primary key of which will be matched to a random integer generated. This system also has a provision of directing the user to applications such as YouTube, Spotify, Indian Express, Kindle, Audible, through which user can be redirected to these apps in the presence of network connection.

Moodyst algorithm
Moodyst (string emotion) labeling the emotion
Matching the label to tables' names in database
Generation of random integer within range of primary key values
Mapping of media and random integer
Playing of media matched

1.3.4 RESULT

This system has been carried out on Android devices, with the given accuracy in following text classification algorithms and feature extraction-based emotion detection algorithms using committee CNNs (Table 1.1).

The following graph shows the frequency of emotions detected in the second dataset (Figures 1.6–1.9).

TABLE 1.1
Accuracy Comparison

Trained Algorithm	Accuracy
Logistic regression	0.7125
Monomial Naive Bayes	0.7183
SVM	0.7829

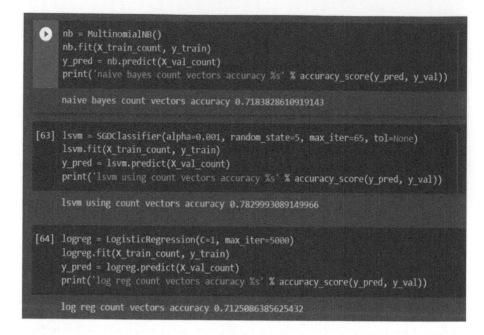

```
nb = MultinomialNB()
nb.fit(X_train_count, y_train)
y_pred = nb.predict(X_val_count)
print('naive bayes count vectors accuracy %s' % accuracy_score(y_pred, y_val))

naive bayes count vectors accuracy 0.7183828610919143

[63] lsvm = SGDClassifier(alpha=0.001, random_state=5, max_iter=65, tol=None)
lsvm.fit(X_train_count, y_train)
y_pred = lsvm.predict(X_val_count)
print('lsvm using count vectors accuracy %s' % accuracy_score(y_pred, y_val))

lsvm using count vectors accuracy 0.7829993089149966

[64] logreg = LogisticRegression(C=1, max_iter=5000)
logreg.fit(X_train_count, y_train)
y_pred = logreg.predict(X_val_count)
print('log reg count vectors accuracy %s' % accuracy_score(y_pred, y_val))

log reg count vectors accuracy 0.7125086385625432
```

FIGURE 1.6   Accuracy of different vectors.

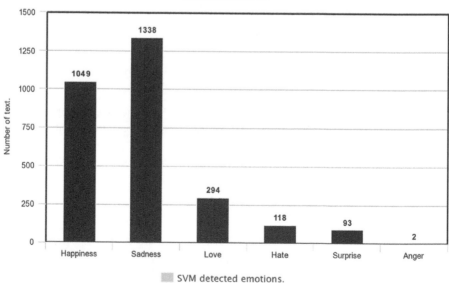

FIGURE 1.7   SVM detected emotions.

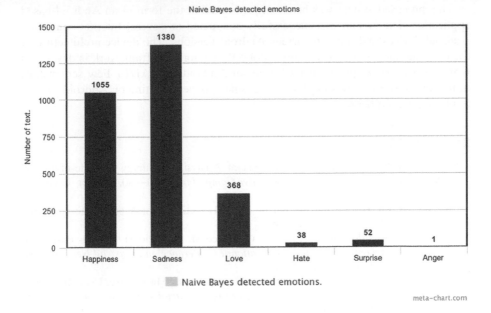

FIGURE 1.8 Naïve Bayes detected emotions.

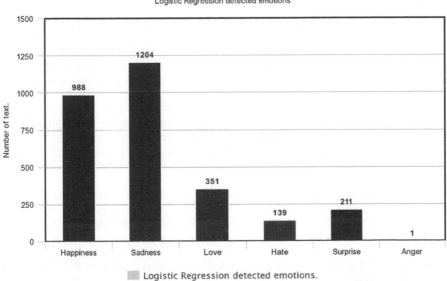

FIGURE 1.9 Logistic Regression detected emotions.

This proposed system was hence implemented in the form of an Android-based application. The system was successfully conducted on several Internet-connected Android devices and runs on various Android versions. The device architecture of the user interface is designed using Adobe Illustrator. The framework is developed with the Android Application Toolkit as an Android app (ADT). Few screenshots of the device from a Samsung F41 model smartphone operating on Android version 6.0.1 are attached below.

REFERENCES

[1] Wang, S, Q Ji, "Video Affective Content Analysis: A Survey of State-of-the-Art Methods," *IEEE Transactions on Affective Computing*, 6, 410–430, October–December 2015.

[2] Liu, D, L Lu, H J Zhang, "Automatic Mood Detection from Acoustic Music Data1," *IEEE*, 8712006, 19 December 2005.

[3] Erion C, Riccardo Coppola Eleonora G, Marco M, Maurizio M, "Mood- Based On-Car Music Recommendations," *International Conference on Industrial Networks and Intelligent Systems*, 154–163, 2016.

[4] Vijayakumar, C, V Adaickalam, "Smart Stress Relieving Music Player Using Intelligent Sentiment Analysis," *International Journal of Industrial Electronics and Electrical Engineering*, ISSN: 2347-6982 4, 6, 2016.

[5] Iyer, A V, V Pasad, S R Sankhe, K Prajapati, "Emotion Based Mood Enhancing Music Recommendation," *IEEE*, 17504486, 5 January 2018.

[6] Kobayashi, H, F Hara, "The Recognition of Basic Facial Expressions by Neural Network," *International Joint Conference on Neural Network*, 460–466, 1991.

[7] Pratama, B Y, R Sarno, "Personality Classification Based on Twitter Text Using Naive Bayes, KNN and SVM," *2015 International Conference on Data and Software Engineering (ICoDSE)*, 170–174, 2015, Yogyakarta, Indonesia: IEEE. DOI: 10.1109/ICODSE.2015.7436992.

[8] Kulkarni, S S, N P Reddy, S I Hariharan, "Facial Expression (Mood) Recognition from Facial Images Using Committee Neural Networks," *BioMedical Engineering OnLine*, 16, 2009.

[9] Kobayashi, H, F Hara, "Recognition of Mixed Facial Expressions by Neural Network," *Proceedings of International Workshop Robot and Human Communication*, 387–391, 1992.

[10] Mariappan, M B, M Suk, B Prabhakaran, "FaceFetch: A User Driven Multimedia Content Recommendation System Based on Facial Expression Recognition," *2012 IEEE International Symposium on Multimedia*, 1, 84–87.

[11] Tripathi, A, T S Ashwin, R M R Guddeti, "WmoWare: A Context-Aware Framework for Personalized Video Recommendation Using Affective Video Sequences," *IEEE Access*, 7, 51185–51200, 2019.

[12] Zhao, J, G Kearney, "Classifying Facial Emotions by Back propagation Neural Networks with Fuzzy Inputs," *Proceedings of Conference on Neural Information Processing*, 1, 454–457, 1996.

[13] Kim, S-M, E Hovy, Determining the sentiment of opinions. In *Proceedings of Coling*, 2004.

[14] Fasel, B, J Luettin, "Automatic Facial Analysis a Survey," *Pattern Recognition*, 36, 259–275, 2003.

[15] Zhang, Z, "Feature-Based Facial Expression Recognition Experiments with a Multi-Layer Perceptron," *International Journal of Pattern Recognition & Artificial Intelligence*, 13, 893–912, 1999.

[16] Kanade T, J Cohn, Y Tian, Comprehensive Database for Facial Expression Analysis. *4th IEEE International Conference on Automatic Face and Gesture Recognition, France*, 2000.

[17] Padgett, C, GW Cottrell. "Representing Face Images for Emotion Classification," *Proceedings of Conference on Advances in Neural Information Processing Systems*, 1996, 894–900.

2 AI- and IoT-Based Body Sensor Networks for Healthcare System
A Systematic Review

Divya Gangwani
Florida Atlantic University

CONTENTS

2.1 INTRODUCTION

Wireless communications along with Internet of Things (IoT) are used in various aspects of our life, such as in the field of medicine, agriculture, entertainment, deep learning technologies to detect fatal conditions [1], Artificial Intelligence (AI) methods to provide smart technological systems [2] and in intelligent home devices. Body sensor network (BSN) is capable of monitoring health-related data of patients without requiring them to visit a medical expert [3]. It is widely useful for people who don't have access to medical provisions and also for elderly people and children. BSN has been proven to be effective in monitoring the health of people and providing accurate results of the diagnosis. IoT based on the healthcare system is also capable of providing physiological information such as electroencephalogram (EEG) and electrocardiography (ECG) through BSN [4]. The healthcare data is collected via sensors and is stored and secured in the cloud, which is useful by the medical experts for scrutinizing the stored data and making predictions or decisions based on the results. This type of intelligent healthcare system [5] is widely used today by medical experts and hospital staff for providing a telemedicine service to patients who can ask questions related to healthcare and consult a medical expert from anywhere in the world [6]. BSN uses the vital signs of the person captured through a wearable sensor device which sends the information to a medical expert through wireless communication technologies as shown in Figure 2.1. It monitors activity such as blood pressure, oxygen levels, body temperature, and cardiac and brain activity. By looking at the current situation, accessing medical advice and hospital administration has become even more difficult and expensive for most people. Using BSN- and IoT-based intelligent healthcare systems provides faster and better services to people so that they can treat and analyze their lifestyle and improve their quality of living. However, this type of intelligent healthcare system collects a large amount of data that is private and confidential [7]. Hence, it is extremely significant to take care of the data security and storage management [8]. It is necessary to protect the data from hacking and external threats [9]. This chapter suggests and analyzes advanced techniques to improve the traditional healthcare system and discusses the latest trends and improvements in IoT-based healthcare systems and provides ways to deal with majority of healthcare

FIGURE 2.1   Wearable devices in BSN.

data using cloud computing methodologies. The main contribution of this chapter is summarized below.

- This chapter highlights the main use of BSN in IoT technology in healthcare systems and other related fields as well.
- We present an architecture of IoT services and BSN technology and briefly explain its importance in the healthcare network.
- We highlight the importance of BSN in IoT technology and provide a summary of how BSN can change the way people think about the healthcare system.
- The chapter demonstrates the deep learning methods to develop a bias-free model which can capture the data received from the BSN signal and output the results by retrieving the data stored on the cloud server.
- We present a detailed survey of current related real-world applications needed to support IoT innovations in BSN technologies for a better healthcare system.
- Finally, we mention the challenges faced in IoT technologies using BSN and demonstrate the methods used to overcome the security issues in the healthcare management system.

Managing a massive amount of healthcare data for enhancing the quality and standard of diagnosis and treatment in patients is always a challenge [10]. Once we solve these challenges, we can achieve an intelligent healthcare system which would provide a better and reliable healthcare service for people.

The remaining section of this chapter is organized as follows: Section 2.2 discusses the literature review of many researchers in BSN and IoT systems. Features of BSN are covered in Section 2.3. Section 2.4 discusses the importance of BSN in IoT. Section 2.5 describes the brief architecture of IoT along with BSN technology. Section 2.6 covers different methods of deep learning models used in developing a BSN application for the healthcare system. Section 2.7 covers the applications of BSN in healthcare as well as in other fields. Section 2.8 discusses the major challenges faced by BSN and provides methods to overcome those challenges. Finally, we conclude the chapter in Section 2.9.

2.2 LITERATURE REVIEW

Several research materials focus on the latest trends and use of wireless BSN and wearable devices. IoT in the healthcare system has changed how the world thinks about the healthcare system and made it better for a lot of people.

Castillejo et al. [11] demonstrated the use of wireless sensor networks in IoT to integrate wearable devices for an e-health application system. The focus was to build an efficient e-healthcare system for women or other people who work in a hazardous environment so that they can report and take necessary action immediately.

He et al. [12] proposed a "wireless body area network (WBAN)" with a focus on the security of medical data. It used integrated circuits and embedded systems to capture data from the WBAN. Even though this chapter has proven to reduce the security issues, it is still a challenge to manage and reduce the security weakness on the client side.

Rana and Bajpayee [13] developed a system to capture patient's data which includes vital signs of the patients such as body temperature, bloody oxygen level, and heart rate, which is stored and maintained in the cloud server to enable easy access for the medical expert from any part of the country. The authors mention the security issues being handled and assess the measures taken for the security breach of the patient data, but there is still a need for a deep study to handle the security issues in different aspects.

Sharma et al. [14] used a star network topology that contains different components in a network to coordinate with the sensor nodes of a "WBAN" that monitors a patient's physiological signs remotely. However, this method is not capable of capturing all the critical signs of the patient which is required in case of an emergency.

Guidi et al. [15] developed a heart failure analysis dashboard system with AI technology to diagnose and analyze the health parameters of the patient and provide a follow-up on time. This dashboard the system provides the most up-to-date information about the patient's health which is useful to the doctors to provide a healthcare plan to the patient.

Hongxu et al. [16] proposed the use of a framework for monitoring healthcare data and providing trustworthy at-home assistance to patients. It uses a data transmission method via servers installed to provide remote healthcare to the patient via medical experts.

Pandey et al. [17] provided an extensive survey on BSN in healthcare IoT. The paper mentions the limitations and challenges faced by these sensors, which should be taken into consideration when developing a BSN architecture but does not provide a systematic framework of how to overcome these challenges to develop a safe and secure BSN in IoT systems.

2.3 FEATURES OF BSN

A BSN system is an important technology that has been used most widely today by many people as it is the most convenient and fastest way to monitor healthcare vitals and keep a track of basic vital signs especially during the Covid-19 pandemic which makes very difficult to visit a doctor [18]. Several features describe the use and accessibility of BSN in IoT which are described in the below paragraph.

- BSN in IoT opens a lot of doors in healthcare systems for patients to discuss their health results with the doctor. It improves the healthcare management system as it collects the data through sensors and analyzes the results of the most critical parts of the body like the brain and heart [19].
- BSN gathers a huge amount of data that is accurate enough for the doctors to make better decisions for their patients. The use of AI in IoT has provided much smarter and improved the outcome of the BSNs in terms of diagnosis and follow-ups.
- BSN helps people from all over the world and provides a huge pool of resources and allows the opportunity to use techniques available in

developed countries for the underprivileged people. AI technology with IoT is not only useful for people but also for doctors or medical professionals as it helps to make better judgments and decisions based on the diagnosis [20].

- BSN with IoT reduces the overall operational cost for people and provides a better outcome and strategy for healthcare management.
- It improves and provides virtualization for in-home services for patients to access medical services and physician assistants in rural areas.
- BSN also has capabilities of providing real-time services with ease of access and security to patients. It guarantees point-to-point reliability through mobile access for faster communication and results.

2.4 IMPORTANCE OF BSN IN IOT

The healthcare technological systems are continuously evolving with time and are becoming more and more advanced with the growing technology. Every part of the company is digitized including the network system, infrastructure, and virtual healthcare. It provides medical care to people at any given place and time, thereby reducing the risk of a patient's life. Sensor networks and IoT have changed the face of the "healthcare system" all over the world; hence, we can say that without the use of BSN technology along with IoT, it is impossible to set up a "patient-oriented" healthcare system.

Sensor and electronic equipment are connected through a network system with the help of BSN as it helps to establish a connection with a patient in a safe environment. The combination of networking and IoT allows the system to transfer the data to any location, thereby reducing the time taken by a person to reach the hospital, stand in the queue for their blood to be drawn, or check the vital stats of a patient like blood pressure, heart rate, and glucose [21].

In today's world, BSN can be built by rearranging the sensors either by positioning them around the body or by placing them in the body. Sensors can also be positioned over the body as a wearable device that collects important information about the person's health. Along with IoT, BSN provides a framework in which human beings can receive proper, "up-to-date", and cutting-edge "technology-driven" healthcare services to enhance their standard of living [22]. These types of services could include medical sickness monitoring, home or hospital appliance management, access to the patient's medical records, automatic contact with a healthcare person in the case of any emergency, and the guidance of available hospitals and staff based on the emergency. However, one of the most difficult challenges in such type of situation is to overcome the network traffic. Managing the system based on an emergency and handling such network traffic can be a huge problem if the resources are limited. Even in BSN, the prototype is such that it includes a variety of sensors and other electronic devices such as routers, servers, LPUs, and sensor nodes. Due to the BSN configuration and its connection, IoT can provide real-time solutions for detecting and treating medical conditions such as malignant cancer, which if not treated on time can spread to other parts of the body as well. Hence, BSN is the main reason which facilitates the transmission of human body information from sensors to other local entity and from local entity to the centralized or personal server.

Apart from this, AI, machine learning technologies, deep learning frameworks, and natural language processing are all parts of BSN, which makes IoT more robust and capable of creating "Electronic Health Records" which will be useful in case of an emergency and would provide help or send messages to the patients' relative in a timely manner [23].

In a nutshell, BSN provides a reliable base to the IoT in a healthcare environment where people can have faith and trust in the system to perform well and provide accurate results so that the person can concentrate on their daily routine without worrying too much about their health. Even then, it is still a challenge when sharing such important information about the medical records of patients. Processing such vital information can cause problems due to network vulnerability which needs to be taken care of to maintain human belief over BSN in the IoT healthcare system.

2.5 ARCHITECTURE OF IOT BASED ON BSN IN HEALTHCARE SYSTEM

The architecture of IoT in BSN consists of three components which constitute the overall working of technology for patients in a secured manner. The components are a BSN, a secured cloud server, and a medical expert who reads or assigns an emergency management system for the patient [24]. The overall architecture and operations of IoT-based BSN are shown in Figure 2.2.

A BSN is the first component in the architecture as shown in Figure 2.2. It is used to capture the vitals of a patient through sensors attached to the human body. These sensors are known as wearable device sensors for example, a smart watch, or it can be sensors attached to the skin on various parts of the body depending on the kind of tests or vital signs being captured by it [25]. Regardless, these sensors can capture real-time information or human vitals such as heart rate, body temperature, blood oxygen levels, blood sugar levels, and humidity through sensors attached to the wearable device. Some sensors which can be attached to the skin can generate

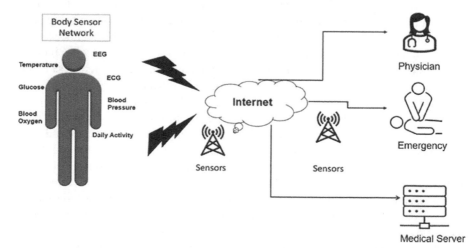

FIGURE 2.2   IoT architecture for the healthcare system.

real-time graphs specifically used for monitoring irregular heartbeats or brain waves through devices known as EEG or ECG [26]. These sensors provide readings of the vitals and display them through visual graphs or nodes for better communication and understanding of the patients. The data generated from these nodes, or the BSN are collected and transmitted via the internet through the cloud servers to the medical experts for further evaluation [27].

The second component is the medical server or cloud service which is majorly responsible for collecting and storing crucial information in the form of data of patient's healthcare. The information is passed on the cloud via the internet with proper security management to prevent unauthentic use of the data and avoid hacking of the servers. These servers also provide the capability to set an alert in the system and notify the patients and the medical experts based on potential health condition of the patient by reading the uploaded data [28]. This capability of the healthcare monitoring system prevents any delays from the patient as well as the doctor's side and ensures timely management of the medical condition.

The third component provides the healthcare management from the medical experts to the patient. Various services are emergency care, hospitalization, on-spot CPR by the nurse or healthcare staff, sending at-home services for elderly patients, and physician visits to the patients who are stable and do not require emergency services depending on their condition [29]. The healthcare management staff coordinates with the patients and monitors their data sent via servers of the medical system so that they can assist the patients as well as doctors and prepare them for emergencies. The healthcare management staff acts as a middleman for patients and doctors. This type of system ensures proper management and coordination between the users and medical experts so that the users can trust the smart technology and can rely on the data generated by the wearable devices [30].

There are many pre-existing architectures of BSN used by many authors based on IoT in the healthcare system. BSN is evaluated on many techniques based on different parameters. The different parameters are techniques are provided in Table 2.1 which shows that the data which is ingested from the sensors is then stored in a secondary server or a remote server to maintain privacy and further analyze the stored medical information to produce the desired outcome as it guarantees further safety of healthcare data and builds trust of users in BSN or medical professionals.

In Table 2.1, we discuss some pre-existing techniques provided by different authors and the parameters taken into consideration such as data security, data management, and ease of access when designing the architecture of the healthcare system.

Many authors have taken data privacy and management into account when designing a BSN architecture; this is because a simple flaw in the design can ruin the main objective of BSN and can also instill fear in the users of the IoT device. Overall, the discussed parameters have been taken into consideration when designing the architecture of BSN, but none of the authors have considered all the parameters together in a unique model. If this can be taken care in the future, then we can say that the modeling of the BSN architecture along with some other parameters can be an ideal solution for a healthcare system as it will fulfill the approach to keep humans safe and less stressed about their health conditions.

TABLE 2.1

Summary of Different Techniques Used in the Past Research of BSNs

Authors	Techniques	Data Security	Data Management	Ease of Access	Energy Efficiency
Stephenie B Baker et al. [31]	Recommendation systems	Not present	Yes	Yes	Yes
Prosanta Gope and Tzonelih Hwang [11]	Decision-based technique with an alert system	Present	Yes	Yes	Yes
Habib Carol et al. [32]	Decision-based technique	Not present	No	Yes	Yes
Alper Akkas et al. [33]	Alert feature	Not present	Yes	Yes	No
Ali Farman et al. [34]	Recommendation technique	Not present	Yes	Yes	No
Bulaghi Zohre et al. [35]	Prediction-based technique	Not present	No	No	No

2.6 AI TECHNOLOGY USING DEEP LEARNING METHODS FOR SENSOR DATA

Many deep learning methodologies have been used in the past which demonstrate effective management and detection of various diseases or illnesses and prevent the condition to get worsened by prolonged delays [36]. It can also provide a faster and accurate diagnosis using sensor data captured from wearable devices as shown in Figure 2.3.

For EEG diagnosis and detection, deep learning technologies such as convolutional neural network have been proposed by [37,38], which captures the brain signals and decodes the information using automatic feature extraction method. This can prevent or detect serious health conditions based on the sensor data. This data is used by the healthcare staff to further diagnose the treatment plan and order more tests such as CT or MRI for the brain to have a clear idea and present the line of treatment to the patient.

The current BSN system with time-series analysis for multiple applications in sports and fitness provides various advantages like self-analyzing techniques, coaching, opponent analysis, tactic analysis, and recruitment. The time-series data can provide real-time information and visualization of live sports events.

For modeling sensor data, a collection of features from the sensor attached to the chest can be represented as:

$$F_f = \left(F_x, F_y, F_z\right) \tag{2.1}$$

Similarly, the sensor attached to the heart for ECG can be represented as:

$$E = (E1, E2) \tag{2.2}$$

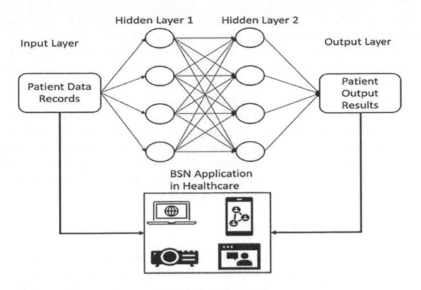

FIGURE 2.3 A deep learning network diagram for BSN.

The sensor attached to the heart for EEG can be represented as:

$$C = (C1, C2) \tag{2.3}$$

Hence, all the features captured by various sensors on the body can be represented as:

$L = F_f \parallel E \parallel C$. These features are then fed to a model for evaluation. To enhance the features a kernel-based discriminant analysis (KDA) is used, which is based on the eigenvalue problem to find the non-linear feature space. This works as follows, the sensors attached to different parts of the body are captured through a Gaussian kernel obtained from the maximization of the following equation:

$$S = \frac{\left| S^T C_w S \right|}{\left| S^T C_B S \right|}$$

where S represents discriminant features of class C_B with class C_w. KDA provides a clustering technique that demonstrates features based on groups of clusters.

Another deep learning methodology known as long short-term memory (LSTM) has been widely used in detection of heart rate monitoring using sensors to detect irregular heart rates in patients. Faust et al. [39] used LSTM to capture signals such as the heart rate to detect atrial fibrillation in patients. A smart monitoring system captures signals or waves generated through sensors used in wearable devices.

A deep neural network (DNN) [40] is used to capture the information from the sensors and use these features to analyze the data and predict the seriousness of the extracted information. A DNN uses hidden layers and the backpropagation method to acquire signals and categorize them based on information or features extracted from the layers of DNN. It consists of three layers: the first layer provides separation of signals using the linear method, the second layer is a "fully connected" layer, and the last layer uses an activation function which is a linear function. These layers build a model for DNN to predict the signals and develop a method for the IoT-based health-care system which provides them with accurate information about the patients. DNN is used for predicting whether the signal is abnormal or normal based on the specific range given by the medical experts to evaluate the signals generated by the device.

The signal logs generate physiological signals like ECG, EEG, body tempera-ture, "heart rate", and "blood oxygen" which are some forms of notable signals that are acquired for the data retrieval process. DNN provides the capability to further analyze only those signals which are abnormal and provides more flexibility for the medical experts to evaluate and take further steps on time.

2.7 APPLICATIONS OF BSN

The BSN is widely studied by many researchers today since BSN along with IoT can change the lives of many people. Its applications are not just limited to the medi-cal field or healthcare management, but it is used in many other fields as well. BSN is used in other fields such as, sports and fitness training, entertainment, disability services as well as in military services [41]. Figure 2.4 shows all the applications of BSN in different fields. The below paragraph briefly describes the most important applications of BSN in the healthcare system.

2.7.1 MEDICAL FIELD

BSN is mostly used in the medical field as it has the widespread capabilities to detect and analyze healthcare data for most of the diseases which affect human lives. The whole world is facing lot of health-related issues and providing timely and effective services to people has been a challenge. Hence, BSN is becoming

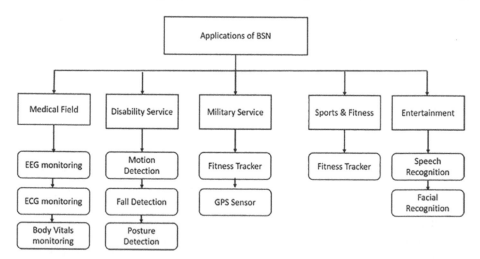

FIGURE 2.4   Applications of BSN in different fields.

increasingly important to detect different types of health issues. Some of them are described below.

ECG signal monitoring is one of the common applications of BSN [42]. It sends signals captured through the sensors to the wearable device which looks for irregular heartbeats or cardiac activity in a patient. ECG monitoring system is very useful for a patient who is at high risk of getting a heart attack as it can diagnose the condition at the correct time and send an alert to the medical experts to handle the situation and provide early diagnosis and treatment [43].

EEG captures the brain waves of the patient through attached sensors on the brain. This type of monitoring system can detect normal or abnormal brain waves and analyze and predict abnormal brain activities such as seizures and notify the medical expert immediately so that they can provide treatment options or send the patient for further diagnosis or scanning of the brain to rule out other possible causes. With these applications and monitoring systems, a patient has a chance to live a healthy life and get treatment on a timely basis [44].

Heart rate, blood pressure, and bloody oxygen monitoring system is widely used and provided in the healthcare monitoring system through a wearable sensor usually worn as a watch or other tracking device which sends the signal to the people in real time.

2.7.2 DISABILITY SERVICES

Many people around the world suffer from disability which could be blindness, learning disability, or paralysis. These types of people often require assistance from someone to complete their day-to-day chores. Especially elderly people who have difficulty in completing their daily activity need a systematic and trustworthy monitoring system to help them live a normal life. WBAN attaches sensors to detect motion and movement control in patients. It also can detect a fall and help them to regain their strength by providing physical activity assistance through the healthcare system. It also helps patients with Parkinson's to detect balance issues and prevents them from falling [45].

A posture detection monitoring system has been used which is equipped with three sensing techniques that are placed on the shoulders to detect motion while movement. This technique captures the user's posture while sitting, standing, or lying down so that it can detect any sudden changes in the movement like tilting or falling [35].

2.7.3 SPORTS AND FITNESS

BSNs are widely used in sports such as swimming, soccer, athletes, and even for people who work out at the gym and want to achieve their fitness goals. BSN is used for training and self-assessment of their performance. These sensors are used in measuring and keeping track of their pace, speed, distance, time, and calories burnt in a day based on the type of activity selected [32].

2.7.4 MILITARY SERVICES

BSN helps in tracking and monitoring the activities of enemies and protects military persons from getting hurt or targeted by enemies. It will improve troop readiness and provide useful alerts to keep them active during the war or attacks. Wireless sensors along with GPS sensors are used to track the movement and provide useful information to military professionals [46]. Apart from this, military professionals use BSN for keeping them active and help them to stay fit throughout the day [33].

2.7.5 ENTERTAINMENT

Nowadays, the entertainment system is ever-changing, and new technologies are being added to entertainment daily be it mobile phones, televisions, or other electronic gadgets. Many latest technologies are going hands-free and require facial movements or speech recognition to operate such devices. This is due to the capability of wireless sensors and IoT technologies that detects motions or face recognition to provide entertainment services with ease of living [47,48].

2.8   CHALLENGES OF BSN AND IOT IN HEALTHCARE TECHNOLOGIES

The key requirements in designing a suitable wireless communication technology based on IoT healthcare system includes a real-time monitoring of the healthcare data, timely management of requests from patients to the medical experts, effective diagnosis and reading of the data, and providing enhanced security of the data. It is extremely important to maintain the privacy and confidentiality of the patient as well as keep the medical data secured.

2.8.1   SCALABILITY

Scalability in the "healthcare system" is an important benchmark that should be taken care of when designing or developing a new application. Scalability is nothing but the capability of any healthcare monitoring system to be able to adapt to the ever-changing environment of healthcare and fulfilling the needs of the users. Due to too many servers, and devices used for healthcare systems via the internet, IoT technologies become less scalable as it loses uniformity and communication techniques. Maintaining uniformity among the devices reduces the scalability issues and makes the use of BSN with IoT more flexible.

The main objective is to make the BSN connect to a device more scalable so that it can meet the needs of the changing environment as scalability ensures the healthcare management system work smoothly without any delays from the server side. Unresponsive devices could cause errors in reading the data and make it less reliable for critical patients to use such devices. Hence, resolving the scalability issues becomes an important factor in the existing devices as well as takes care of scalability before developing a new device for future use.

2.8.2   SECURITY

In today's world, as the data keep increasing day by day, it is imperative to take necessary steps with the aim to improve the security and privacy of the "healthcare monitoring system".

Researchers and engineers must ensure that the healthcare monitoring systems that are used for capturing various applications have authenticated security as the use of these applications is increasing with time [34].

Keeping patient confidentiality and the importance of the medical data in mind, the healthcare system should take steps to improve the security, maintain confidentiality as well as ensure data quality.

Based on the IoT healthcare environment, the "smart healthcare application system" is one of the most important applications for physicians to monitor the state of corresponding patients remotely via WBAN [49]. When information is exchanged through networks, phishing scams are one of the harmful attacks that might be expected. Hence, measures should be in place when designing a smart healthcare system.

Gope et al. [50] introduced a method known as BSN-care which provides enhanced security to the patients in the healthcare application system. The total security framework was divided into two parts: network security and data security. Authentication, localization, and data anonymization are all aspects of network security. On the other hand, data privacy and integrity are handled by data security. Together these two components provide overall security and authentication of the healthcare system and prevent the data from forgery attacks.

Lal et al. [35] proposed a "content-centric network" method based on "WBAN" for IoT healthcare. This method provides high security in applications that require connectivity to portals or another form of means to access patient data. To reach the patient information to the doctor, it goes through a lot of people who serve as the middlemen between the patient and doctor; hence, it is important to make sure the information is not accessible by wrong people and ensure seamless connectivity, and high data transmission rate at a faster speed provides better access to the information.

Wu et al. [51] used an anonymous authentication system for maintaining enhanced security inpatient data information in healthcare applications. This method reduces the security issues in the authentication method which is most widely faced by many people. The author was able to achieve high security with better performance at a reduced cost in the healthcare system.

2.8.3 PRIVACY AND CONFIDENTIALITY

Privacy and confidentiality in a healthcare system are important for both patient and the doctor. It keeps the relationship between the patient and the doctor fair and builds trust in each other. Only a person's doctor, healthcare provider, or health insurance provider should have access to such sensitive and confidential information. Any misuse or misconduct would lead to disastrous consequences in the medical system [52]. A lawsuit can be filed against the medical supervisor or the health insurance company due to a data privacy breach. Hence, the healthcare system should be safe, private, and must have enhanced security measures to keep the data confidential [53]. Therefore, BSN in IoT technology should overcome security and privacy issues as well as provide a faster and reliable source for patients and doctors to trust the system and improve their lifestyle [42].

2.9 CONCLUSION

This chapter summarizes the use of wearable BSN sensors to detect signals related to healthcare in patients. There has been a significant amount of research over the last few years in the field of smart healthcare systems with the use of BSN or WBAN. Today's healthcare industry is arguably the most vital and delicate sector as it contributes to a healthy lifestyle in people. If the healthcare system is not properly managed, it can harm a person's emotional and social well-being, and even put their lives in jeopardy. Hence, it is important to take measures in providing the utmost security and high-end technology models in use. In this chapter, we summarize the use of the deep learning model for building a smart healthcare system. We highlight the architecture of IoT in healthcare and focus on providing various applications and use of BSN in healthcare and highlight the use in other fields as well. BSN has shown capabilities to monitor healthcare data remotely, which is a major boom for elderly people.

Studies have shown how BSN can change the lives of elderly as well as young people. The advanced capabilities and smart monitoring devices with ease of access to wearable sensors deliver promising and trustworthy results and ensure a good quality of life for people. However, with such a critical system, there is always the concern with privacy and security risks involving patients' data. With proper management of data and providing high-end security, the BSN with IoT can provide a huge change in the quality of life of people and can be the future of the medical revolution in the future.

In the upcoming years, the applications of BSN can include personalized prescriptions and medicines delivered at home and send reminders about the refills or expiry of the medicines. It also has the capability of extending its applications in other fields that have not been used before. Hence, BSN in IoT with AI technology provides revolutionizing change to the society for the people as well as the medical experts.

REFERENCES

[1] N Al Bassam, S Asif Hussain, A Al Qaraghuli, J Khan, E P Sumesh, and V Lavanya. IoT-based wearable device to monitor the signs of quarantined remote patients of covid-19. *Informatics in Medicine Unlocked*, 24: 100588, 2021.

[2] H N Almajed, A S Almogren, and A Altameem. A resilient smart body sensor network through pyramid interconnection. *IEEE Access*, 7: 51039–51046, 2019.

[3] P Castillejo, J-F Martinez, J Rodriguez-Molina, and A Cuerva. Integration of wearable devices in a wireless sensor network for an e-health application. *IEEE Wireless Communications*, 20, 4: 38–49, 2013.

[4] W-Y Chung, Y-D Lee, and S-J Jung. A wireless sensor network compatible wearable u-healthcare monitoring system using integrated ECG, accelerometer, and spO2. In *2008 30th Annual International Conference of the IEEE Engineering in Medicine and Biology Society*, 1529–1532. IEEE, 2008.

[5] D Gangwani, Q Liang, S Wang, and X Zhu, An empirical study of deep learning frameworks for melanoma cancer detection using transfer learning and data augmentation, *2021 IEEE International Conference on Big Knowledge (ICBK)*, 2021, 38–45, doi: 10.1109/ICKG52313.2021.00015.

[6] L D'Arco, H Zheng, and H Wang. Sensebot: A wearable sensor-enabled robotic system to support health and well-being. *CERC*, 30–45, 2020.

[7] O Faust, A Shenfield, M Kareem, T R San, H Fujita, and U Rajendra Acharya. Automated detection of atrial fibrillation using long short-term memory network with r interval signals. *Computers in Biology and Medicine*, 102: 327–335, 2018.

[8] P Gangwani, A Perez-Pons, T Bhardwaj, H Upadhyay, S Joshi, and L Lagos. Securing environmental IoT data using masked authentication messaging protocol in a DAG-based blockchain: IOTA tangle. *Future Internet*, 13, 12: 312, 2021. https://doi.org/10.3390/fi13120312.

[9] H Fouad, A S Hassanein, A M Soliman, and H Al-Feel. Analyzing patient health information based on IoT sensors with AI for improving patient assistance in the future direction. *Measurement*, 159: 107757, 2020.

[10] D Gangwani and P Gangwani. Applications of machine learning and artificial intelligence in intelligent transportation system: A review. *Applications of Artificial Intelligence and Machine Learning*, 203–216, 2021.

[11] P Gangwani, J Soni, H Upadhyay, and S Joshi. A deep learning approach for modeling of geothermal energy prediction. *International Journal of Computer Science and Information Security (IJCSIS)*, 18, 1, 2020.

[12] P Gope and T Hwang. Bsn-care: A secure IoT-based modern healthcare system using body sensor network. *IEEE Sensors Journal*, 16, 5: 1368–1376, 2015.

[13] L Greco, G Percannella, P Ritrovato, F Tortorella, and M Vento. Trends in IoT based solutions for health care: Moving into the edge. *Pattern Recognition Letters*, 135: 346–353, 2020.

[14] G Guidi, M Chiara Pettenati, R Miniati, and E Iadanza. Heart failure analysis dashboard for patient's remote monitoring combining multiple artificial intelligence technologies. In *2012 Annual International Conference of the IEEE Engineering in Medicine and Biology Society*, 2210–2213. IEEE, 2012.

[15] K Hameed, I Sarwar Bajwa, S Ramzan, W Anwar, and A Khan. An intelligent IoT based healthcare system using fuzzy neural networks. *Scientific Programming*, 2020, 2020.

[16] A Harbouche, N Djedi, M Erradi, J Ben-Othman, and A Kobbane. Model driven flexible design of a wireless body sensor network for health monitoring. *Computer Networks*, 129: 548–571, 2017.

[17] D He, S Zeadally, N Kumar, and J-H Lee. Anonymous authentication for wireless body area networks with provable security. *IEEE Systems Journal*, 11, 4: 2590–2601, 2016.

[18] A Khanna, M Kapahi, and E Gupta. Wearable technologies in IoT: A critical survey. In: Proceedings of the International Conference on Innovative Computing & Communication (ICICC). Available at SSRN 3833933, 2021.

[19] X Lai, Q Liu, X Wei, W Wang, G Zhou, and G Han. A survey of body sensor networks. *Sensors*, 13, 5: 5406–5447, 2013.

[20] K Nidhi Lal and Anoj Kumar. E-health application over 5g using content-centric networking (CCN). In *2017 International Conference on IOT and Application (ICIOT)*, 1–5. IEEE, 2017.

[21] R Kumar Mahendran and P Velusamy. A secure fuzzy extractor based biometric key authentication scheme for body sensor network in internet of medical things. *Computer Communications*, 153: 545–552, 2020.

[22] V Mainanwal, M Gupta, and S Kumar Upadhayay. A survey on wireless body area network: Security technology and its design methodology issue. In *2015 International Conference on Innovations in Information, Embedded and Communication Systems (ICIIECS)*, 1–5. IEEE, 2015.

[23] H Masoumi, A Behrad, M Ali Pourmina, and A Roosta. Automatic liver segmentation in MRI images using an iterative watershed algorithm and artificial neural network. *Biomedical Signal Processing and Control*, 7, 5: 429–437, 2012.

[24] B Anand Muthu, CB Sivaparthipan, G Manogaran, R Sundarasekar, S Kadry, A Shanthini, and A Dasel. IoT based wearable sensor for diseases prediction and symptom analysis in healthcare sector. *Peer-to-peer Networking and Applications*, 13, 6: 2123–2134, 2020.

[25] M Muzammal, R Talat, A Hassan Sodhro, and S Pirbhulal. A multi-sensor data fusion enabled ensemble approach for medical data from body sensor networks. *Information Fusion*, 53: 155–164, 2020.

[26] A Nadeem, M Azhar Hussain, O Owais, A Salam, S Iqbal, and K Ahsan. Application specific study, analysis and classification of body area wireless sensor network applications. *Computer Networks*, 83: 363–380, 2015.

[27] MA Kumar and YR Sekhar. Android based health care monitoring system. In 2015 International Conference on Innovations in Information, Embedded and Communication Systems (ICIIECS). IEEE, 2015.

[28] C Pandey, S Sharma, and P Matta. Body sensor network architectures in healthcare internet-of-things (HIoT): A survey. In *2021 6th International Conference on Communication and Electronics Systems (ICCES)*, 494–499. IEEE, 2021.

[29] H Kupwade Patil and R Seshadri. Big data security and privacy issues in healthcare. In *2014 IEEE International Congress on Big Data*, 762–765. IEEE, 2014.

[30] S Furqan Qadri, S Afsar Awan, M Amjad, M Anwar, and S Shehzad. Applications, challenges, security of wireless body area networks (WBANS) and functionality of IEEE 802.15.4/zigbee. *Science International*, 25, 4: 697–702, 2013.

[31] H Yin and N K Jha. A health decision support system for disease diagnosis based on wearable medical sensors and machine learning ensembles. IEEE Transactions on Multi-Scale Computing Systems, 3, 4: 228–241, 2017.

[32] S B Baker, W Xiang, and I Atkinson. Internet of things for smart healthcare: Technologies challenges and opportunities. *IEEE Access*, 5: 26521–26544, 2017.

[33] H Carol, A Makhoul, R Darazi, and R Couturier. Health risk assessment and decision-making for patient monitoring and decision-support using wireless body sensor networks. *Information Fusion*, 47: 10–22, 2019.

[34] M Alper Akkas, R Sokullu, and H Ertürk Cetin. Healthcare and patient monitoring using IoT. *Internet of Things*, 11, 2020.

[35] S Sanjay Kale and D S Bhagwat. A secured IoT based web care healthcare controlling system using BSN. In *2018 Second International Conference on Inventive Communication and Computational Technologies (ICICCT)*, 816–821, 2018.

[36] P Rajan Jeyaraj and E Rajan Samuel Nadar. Smart-monitor: Patient monitoring system for IoT-based healthcare system using deep learning. *IETE Journal of Research*, 68, 1435–1442 2019.

[37] J Rana and A Bajpayee. Healthcare monitoring and alerting system using cloud computing. *International Journal on Recent and Innovation Trends in Computing and Communication*, 3, 2: 102–105, 2015.

[38] D Ravi, C Wong, B Lo, and G-Z Yang. A deep learning approach to on-node sensor data analytics for mobile or wearable devices. *IEEE Journal of Biomedical and Health Informatics*, 21, 1: 56–64, 2016.

[39] P P Reboucas Filho, R Moura Sarmento, G Bandeira Holanda, and D de Alencar Lima. New approach to detect and classify stroke in skull CT images via analysis of brain tissue densities. *Computer Methods and Programs in Biomedicine*, 148: 27–43, 2017.

[40] SD Ross, R Estok, S Chopra, and J French. Management of newly diagnosed patients with epilepsy: A systematic review of the literature. *Evidence Report/Technology Assessment (Summary)*, 39: 1–3, 2001.

[41] S Sharma, A Lal Vyas, B Thakker, D Mulvaney, and S Datta. Wireless body area network for health monitoring. In *2011 4th International Conference on Biomedical Engineering and Informatics (BMEI)*, 4: 2183–2186. IEEE, 2011.

[42] H F Nweke, Y W Teh, M A Al-Garadi, and U R Alo. Deep learning algorithms for human activity recognition using mobile and wearable sensor networks: State of the art and research challenges. *Expert Systems with Applications*, 105: 233–261, 2018.

[43] Y Sun. *Securing Body Sensor Networks and Pervasive Healthcare Systems*, Imperial College, London, 2019.

[44] C A Tavera, J H Ortiz, O I Khalaf, D F Saavedra, and THH Aldhyani. Wearable wireless body area networks for medical applications. *Computational and Mathematical Methods in Medicine*, pp. 1–9, 2021.

[45] L Wu, Y Zhang, L Li, and J Shen. Efficient and anonymous authentication scheme for wireless body area networks. *Journal of Medical Systems*, 40, 6: 1–2, 134, 2016.

[46] S Subrata, K Singh, S Das, and R Kumar. Energy efficient shortest path routing protocol in underwater sensor networks. In *2016 International Conference on Computing, Communication and Automation (ICCCA)*, 546–550. IEEE, 2016.

[47] S Vivek Kumar, R Kumar, and S Sahana. To enhance the reliability and energy efficiency of WSN using new clustering approach. In *2017 International Conference on Computing, Communication and Automation (ICCCA)*, 488–493. IEEE, 2017.

[48] H A Sanghvi, S B Pandya, P Chattopadhyay, R H Patel, and A S Pandya. Data science for e-healthcare, entertainment and finance. In *2021 Third International Conference on Inventive Research in Computing Applications (ICIRCA)*, 604–611. IEEE, 2021.

[49] M M Dhanvijay, and S C Patil. Internet of Things: A survey of enabling technologies in healthcare and its applications. *Computer Networks*, 153: 113–131, 2019.

[50] A Farman, S El-Sappagh, S M Riazul Islam, D Kwak, A Ali, M Imran, et al. A smart healthcare monitoring system for heart disease prediction based on ensemble deep learning and feature fusion, *Information Fusion*, 63: 208–222, 2020.

[51] B Zohre Arabi, A Habibi Zad Navin, M Hossein Zadeh and A Rezaee. SENET: A novel architecture for IoT-based body sensor networks, *Informatics in Medicine Unlocked*, 20, 100365, 2020.

[52] C Pandey, S Sharma, and P Matta. Body sensor network architectures in healthcare Internet-of-Things (HIoT): A survey. In *2021 6th International Conference on Communication and Electronics Systems (ICCES)*, 494–499, 2021. doi: 10.1109/ICCES51350.2021.9489205.

[53] A A Mutlag, M K A Ghani, and M A Mohammed. A healthcare resource management optimization framework for ECG biomedical sensors. In *Efficient Data Handling for Massive Internet of Medical Things*, 229–244. Springer, Cham.

3 On the Convergence of Blockchain and IoT for Enhanced Security

Pranav Gangwani
Florida International University

Tushar Bhardwaj, Alexander Perez-Pons,
Himanshu Upadhyay, and Leonel Lagos
Florida International University

CONTENTS

3.1 INTRODUCTION

Internet of Things (IoT) is a global network of intelligent physical devices (referred to as "Things") [1] and people. The IoT allows every "Thing" to link and connect, effectively transforming the entire physical ecosystem into a massive data managing environment. Cloud computing [2], artificial intelligence [3], and deep learning [4], as well as analyzing data and modeling of information, are all turning out to be important components of the IoT system. The great progress in the domain of IoT is also creating development in the "Information and Communication Technology" industry.

DOI: 10.1201/9781003248750-3

Smart devices gather data from their local environment and then communicate with other platforms and IoT devices. Data is gathered and stored on a centralized server in traditional IoT networks for later use (particularly cloud servers) [5]. As a result, IoT consumers must develop trust in the traditional central servers, knowing that the confidential and sensitive information is protected there. Despite the undeniable benefits that the service providers deliver, centralized IoT systems sometimes confront some obstacles. Unencrypted server data, for example, can be hacked, resulting in the disclosure of important information. A booming technology called blockchain is evolving which is a ledger that is distributed and immutable. This technology works by recording and storing transactions using its cryptographic algorithms and shared database. The participants in blockchain operate in a peer-to-peer manner on the Internet. Technically, blockchain resolves trust-related issues. The hash-chain-based encryption algorithm's safe transmission with certificate value's time stamp mechanism provides data traceability and irreversible change.

In this book chapter, various security risks of IoT will be discussed such as issues with aggregated IoT data which can happen if the sensors are not monitored before aggregating the data. Some features of IoT can also pose a threat such as data storing, data transmission, embedded tag, IoT infrastructure, authentication [6], and access control. Furthermore, we will discuss why we need blockchain and an overview of the distributed ledger technology. Moreover, we will discuss the combination of blockchain with the IoT to enhance the security of IoT devices. Furthermore, we will discuss the benefits of the integration of these two emerging technologies such as decentralization, security, scalability, reliability, identity, autonomy, secure code deployment, and market of services. We will also present the issues/challenges that must be overcome when blockchain and IoT are integrated.

Section 3.2 of the book chapter discusses the motivation and the need for blockchain. In Section 3.3, we talk about the IoT architecture and the various security risks that need to be addressed. Section 3.4 elaborates on the background concepts of blockchain technology. Section 3.5 talks about the blockchain-IoT approach which integrates the two core technologies. Section 3.6 presents the challenges that come up when blockchain is integrated with IoT. Finally, Section 3.7 concludes the book chapter.

3.2 MOTIVATION

IoT is a field that is growing at an alarming rate and consists of billions of connected devices worldwide. These devices will continue to increase in the future, and by 2025, this number is expected to reach 75 billion [7]. Data is shared by each of these devices on the Internet. Therefore, it is fairly understood that data will be produced continuously at a huge volume when these massive number of devices are considered. Hence, solving the security problems for such a massive amount of data becomes a major challenge. In this section, we will explain several IoT challenges and why blockchain is required.

The distributed architecture of IoT is one of the most significant concerns. Each node in an IoT cluster could be used to execute "Distributed Denial-of-Service (DDoS)" and many other cyber-attacks, thus making it a potential failure point [8]. An IoT device cluster can quickly collapse if several devices working simultaneously are compromised. Another issue with IoT is its centralized architecture which can become a central point of failure [9], and hence, it must be addressed. Data authentication and confidentiality [10] is another critical and one of the most important threats that must be addressed. If an IoT device is compromised, the data can be misused in many ways. Additionally, security of IoT data becomes crucial [11] in scenarios where the data, computational power, or electricity is being shared by devices autonomously in various application domains.

3.3   IOT SECURITY

3.3.1   IoT Background

The IoT stack contains four layers, namely, sensing, network, middleware, and application layer [12] which is portrayed in Figure 3.1. The sensing layer contains various IoT devices such as various categories of sensors, actuators, QR code scanners, and GPS modules. If an industry scenario is considered, these devices might be attached with some equipment such as conveyor systems, automated guided vehicles, and robotic arms. The network layer contains various technologies such as Bluetooth, MQ Telemetry Transport (MQTT), Wi-Fi, Zigbee, IPv6, and IPv4. All these protocols have the same task, i.e., to transfer data to the next layer for processing and analysis. The middleware layer contains data processing systems such as databases, servers which are responsible for providing storage and computation services such as computing optimization algorithms, decision-making, and maintain records of the huge volume of data. The final layer, i.e., the application layer, delivers the specific needs for various applications to the end users. Supply chain [13], digital healthcare

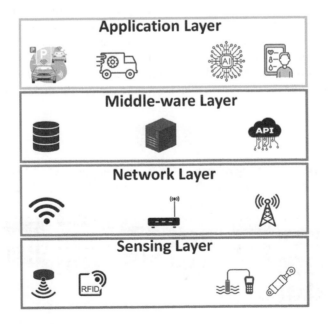

FIGURE 3.1   IoT architecture.

[14], traffic monitoring, smart cities, self-driven cars, etc. are some of the examples of applications that are enabled by IoT.

3.3.2 IoT Security Risks

As discussed in Section 3.3.1, we explained the architecture of IoT which consists of four layers and the responsibilities of each layer in the entire system. Several security issues and threats may arise due to the diverse technologies being used in these four layers. In this section, we will explain the various threats to security pertaining to each of the four layers in the IoT stack as displayed in Figure 3.2.

3.3.2.1 Security Risks at Sensing Layer

The sensing layer deals with various physical IoT devices, for example, sensors and actuators. The sensors' role is to sense or collect the physical surrounding information. On the other hand, based on the sensory data, the actuators execute certain actions on the physical environment. Temperature sensors, vibration sensors, gas sensors, smoke detection sensors, etc. are some of the different types of sensors which are utilized to sense distinct types of data [15]. Various applications of IoT utilize different technologies such as GPS, Robust Security Network (RSNs), radio-frequency identification (RFID), and wireless sensor networks (WSNs). The following are the security risks related to the sensing layer:

1. **Node Capture Attacks:** Sensors and actuators are some of the low-powered nodes which are used in various IoT applications. These low-powered devices are susceptible to a range of threats by any adversary. There might be a possibility that the adversary can replace or capture the node with a compromised node in the IoT system. The replaced node is controlled by the adversary and may appear to adapt with the IoT system. Therefore, the complete security of the IoT application can be compromised [16].
2. **False Data Injection Attack:** An adversary can insert false or invalid information inside the IoT system if an attempt to capture the node is successful.

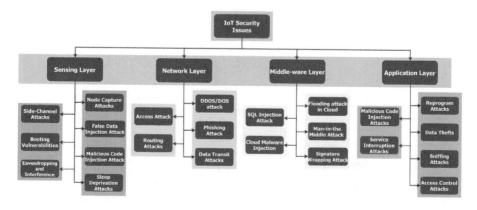

FIGURE 3.2   Threats to IoT security.

This might lead to malfunctioning of the IoT application and may lead to incorrect results as well. An attempt to launch a DDoS attack can also be possible by the attacker if this method is used [17].

3. **Malicious Code Injection Attack:** In this form of threat, the attacker inserts malicious script into the memory of the node. Commonly, the upgrades of the software or firmware are done over the air, which provides an opportunity which can be used by adversaries to infuse a program which includes malware. These adversaries might be able to access the entire IoT system or may even force the nodes to execute certain inessential functions by injecting such malicious code [18].

4. **Sleep Deprivation Attack:** The power source of the low-powered IoT devices is drained by the attackers in this type of attack. Since the battery becomes dead, a denial of service in the IoT application can be launched originated from the nodes. Sleep deprivation attacks could be performed if an adversary can artificially increase the power consumption of various IoT devices or if the attacker is successful in using malicious code to run infinite loops in these IoT devices [19].

5. **Side-Channel Attacks:** Attackers can leak sensitive and confidential data by launching different side-channel attacks contrary to the direct attack on the nodes. Sensitive data can be revealed to the attackers due to the power consumption and the electromagnetic emanation of processors and their microarchitectures. Electromagnetic attacks, laser-based attacks, power consumption and timing attacks are some of the types of attacks on which side-channel attacks are based upon. Cryptographic modules are implemented by modern chips which provide different methods to avoid these attacks [20].

6. **Booting Attacks:** IoT devices become vulnerable to various threats during the boot process. The vulnerability occurs due to the disabled inbuilt security processes at that instance. Therefore, adversaries may take benefit of this susceptibility and may attack the devices while they are in the process of booting. It is necessary to secure the boot process of the edge devices since they are low powered and sometimes undergo sleep-wake cycles [21].

7. **Eavesdropping and Interference:** There are various nodes deployed in open environments that are contained within the IoT applications. Therefore, these IoT applications are at a risk of exposure to eavesdroppers. Throughout different stages such as authentication of data or transmission, the adversaries might be able to capture or eavesdrop the data [22].

3.3.2.2 Security Risks at Network Layer

The network layer's role is to transfer information from the sensing layer to the middleware layer or process unit for further analysis. The following are the security concerns at the network layer:

1. **DoS/DDoS Attack:** In these types of threats, a huge quantity of unwanted requests is executed by an adversary to flood the target servers. Therefore, services to genuine users are disrupted as the target server is incapacitated.

The attack is characterized as a DDoS attack when an attacker employs many sources to overload the target server. Because of the diversity and difficulty of IoT networks, the network layer of the IoT stack is susceptible to these threats. If the IoT devices present in the IoT applications are not correctly installed, attackers can launch DDoS attacks on the target servers, which can serve as simple gateways for them [23].

2. **Phishing Attack:** Phishing attacks occur when an attacker employs minimal amount of effort to target a huge quantity of IoT devices. An expectation of the attacker is that at least some devices will get compromised due to the attack. While visiting web sites on the Internet, there may be a possibility to confront phishing sites by the users. The entire system of IoT which is currently utilized by a user can become susceptible to cyber-threats if the user's login credentials are compromised. In the IoT architecture, the network layer is extremely vulnerable to phishing attacks [24].

3. **Data Transit Attacks:** There is a lot of data exchange and storage involved in various IoT applications. Adversaries and hackers always attempt to target the data since it is valuable. When information is stored on the cloud or on local servers, there is a risk of security breach; however, when it is moved from one location to another, it becomes more vulnerable to cyber-threats [25].

4. **Access Attack:** This is a type of attack where an adversary or any unapproved user achieves access to the IoT network and is also known as advanced persistent threat. The intruder if wishes to can remain inside the network undetected for a long duration. Since applications of IoT constantly transmit and receive vital information, they are especially vulnerable to these sorts of attacks [26].

5. **Routing Attacks:** These attacks work because while data is in transit in an IoT application, malicious nodes may try to reroute the routing pathways. A type of routing attack in which an attacker uses advertisement to encourage nodes to route traffic along the shortest path possible is known as Sinkhole attack. Another attack known as wormhole attack if combined with sinkhole attacks or other kinds of attack can become a serious security threat. The purpose of a wormhole is for fast transfer of packets and is an out-of-band link between two nodes [27].

3.3.2.3 Security Risks at Middleware Layer

In the IoT stack, the middleware layer serves as an abstraction layer in between the network and application layers. The demands of the application layer are fulfilled as this layer provides APIs. The middleware layer is prone to various attacks even though it is useful and provides a robust and reliable IoT application. The following are the different attacks in the middleware:

1. **Flooding Attack in Cloud:** In this type of threat, the quality of service is affected and is similar to Denial-of-Service (DoS) attack in the cloud. The adversaries consistently send numerous requests to a cloud service to deplete the cloud resources. If the load on the cloud servers is increased,

then these attacks could create a huge effect on the resources which are deployed on the cloud [28].

2. **Man-in-the-Middle Attack:** Consider the example of the MQTT protocol which utilizes a communication model in the form of publishing and subscribing. There are three major roles in this protocol known as publisher (client), subscriber, and MQTT broker which efficiently acts as a proxy. Due to the presence of the MQTT broker, messages can be sent without any awareness about the destination and assists to decouple the publisher and subscriber from each other. The attacker could become a man-in-the-middle if the broker is managed by the attacker and full control of all the communication is acquired where the clients are not aware.

3. **Signature Wrapping Attack:** Inside the web services that are utilized in the middleware, XML signatures are employed. The attacker in a signature wrapping attack is able to implement or alter eavesdropped data by manipulating the susceptibilities in Simple Object Access Control unless the adversary is successful in defeating the signature scheme.

4. **SQL Injection (SQLI) Attack:** The middleware layer is prone to another type of attack known as SQLI attacks. Injection of malicious SQL statements can be done by the attacker in this type of attack. After the injection, the attackers can alter records in a database or may even obtain the confidential data of any user.

5. **Cloud Malware Injection:** The attack is known as cloud malware injection if an adversary is able to gain control by injecting malicious script or a virtual machine into the cloud. A malicious service module or a virtual machine is created by the attacker while pretending to be a valid user. As a result, the attacker can get a hold of the victim's service requests and private information that could be altered depending on the situation.

3.3.2.4 Security Risks at Application Layer

The application layer's role is to deliver functional services to customers, and it's where IoT applications like wearables, digital healthcare, and smart retail exist. There are specific issues related to security such as privacy of data and data theft in this layer that are not present in other layers. There are even specific security issues for specific IoT applications. The various security threats to the application layer are explained as follows:

1. **Reprogram Attacks:** The adversary, in these types of attacks, attempts to reprogram the IoT devices remotely if the process of programing is not protected and ultimately leads to the hijacking of the IoT network [29].

2. **Data Thefts:** A vast amount of confidential and sensitive data is dealt by various IoT applications. The data at rest or the static data is less vulnerable than the data in transit and frequent data movements occur in applications of IoT. When such IoT applications are susceptible to data theft threats, the clients will be hesitant to store or use their personal data on these IoT applications.

3. **Sniffing Attacks:** In various IoT applications, sniffer applications may be used by attackers to examine the packets of network traffic. Due to this, the attacker can gain control of the private user information if there are not sufficient security protocols implemented to avert the attack [30].

4. **Access Control Attacks:** Access control is a way for granting or denying access to the account or data to only authorized processes or individuals. If permission control is compromised, the complete IoT application can become prone to threats; hence, an access control attack is a critical threat in IoT applications.

5. **Malicious Code Injection Attacks:** An attacker could invade into a network or a system by commonly opting for the simplest or the most convenient method they can use. Due to insufficient code checks, the system can become vulnerable to malicious scripts and that could become the first point of entry that the attacker would opt. Cross-site scripting is commonly used by attackers to insert malicious code into a trusted website.

6. **Service Interruption Attacks:** According to existing literature, these attacks are also known as DDoS attacks or illegal interruption attacks and many IoT applications have been a victim of this attack in various instances. These attacks artificially make the servers, or a network, too engaged to respond, and thus, the authorized users are unable to use the services of the IoT application [31].

3.4 BLOCKCHAIN TECHNOLOGY

Blockchain technology has been proven to be immutable which is based on the digital ledgers in a distributed manner without the use of any centralized repository and authority to manage and govern it. At the initial phase, blockchain enables the user domain to have transactions in terms of the shared ledger under the umbrella of user domain [32]. The basic idea is that during the normal operations of the blockchain network, the transactions remain unchanged once published.

Let's dig deep into the architecture of blockchain technology which is shown in Figure 3.3. The backbone of blockchain are the digitally distributed ledgers which are signed cryptographically and finally grouped into the blocks [33]. Each of these blocks is then attached/linked to its predecessor block which makes in tamper proof by leveraging the validation mechanism [34] and a consensus decision. Furthermore, when an additional block is appended to the chain, changing the preceding blocks becomes extremely complex (which makes it tamper-resistant).

3.5 INTEGRATION OF BLOCKCHAIN WITH IOT

We describe how blockchain technology can be used where the data about the access control is distributed and stored by proposing an architecture which discusses a decentralized system for access management. All the nodes except the IoT devices and the management hub will be included in the blockchain network. Most IoT devices will be unable to store blockchain data due to the constraints of low-powered IoT devices. We also design a specific node known as the management hub, which

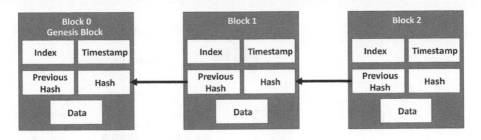

FIGURE 3.3 Blockchain structure.

has the capacity to request data from the blockchain on behalf of IoT devices, such as access control information [35]. Furthermore, the architecture employs a self-executing script called smart contract; this smart contract acts as a solution in which the access control operations can be defined. Special users known as managers define the policy for the smart contract by interacting with it.

3.5.1 Architecture of the System

Figure 3.4 shows the full system architecture which is divided into six various components:

1. Management hub
2. WSNs
3. Blockchain
4. Agent node
5. Managers
6. Smart contract

1. **Management Hub:** As previously stated, the IoT devices in the system design will not be present in the blockchain network. The majority of IoT devices have low power and storage, as well as a limited CPU. Because of these constraints, IoT devices find it difficult to join in the blockchain network directly. We introduced the management hub, a unique node that serves as an interface whose role is to translate data from the IoT devices which are encoded in constrained application protocol messages into JavaScript Object Notation Remote Procedure Call data which can be readable by blockchain participants. A node in the blockchain such as a miner is directly linked with the management hub. As shown in Figure 3.4, a node in the blockchain can have many management hubs linked to it and similarly a management hub can have many WSNs linked to it. IoT devices may request data from the blockchain via the management hub.

    The management hub nodes cannot have any limits comparable to IoT devices since they must be able to handle many concurrent queries from the IoT devices and, hence, require high performance.

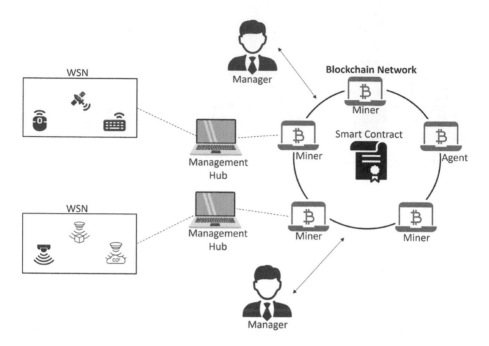

FIGURE 3.4   IoT-blockchain–integrated architecture.

Only in a particular circumstance when authentication is not necessary, any IoT device can contact the blockchain network if it is directly connected to a management center. Although access control may be necessary in several situations, if this is the case, IoT devices would only connect to certain management hubs. The manager node of a specific IoT device must inform about the credentials of that IoT device to the particular management hub once that object is added to the network. The manager node's job also includes informing the IoT device about the management hub's position.

2. **WSNs:** WSNs are special types of transmission networks which provide various applications with low requirements, restricted power, and permits limited connectivity. Additionally, the WSN contains IoT devices which have certain limitations in terms of availability of energy, memory, and power of computation.

The network of the blockchain does not contain IoT devices, and there exists a requirement in our system that all the IoT devices must be globally identified by their unique identity in the blockchain platform. Production of unique and considerably large random numbers can become a problem; however, the usage of public key generators, on the other hand, might be a viable alternative. Usually, a public key for each IoT device would be automatically generated due to the current techniques of cryptography. Therefore, unique identities can be generated by employing various encryption technologies.

3. **Blockchain:** We will use a private blockchain [36] for our architectural system to attain better performance in managing the IoT devices. Since the private blockchain is more scalable than the public blockchain, we opted for that. Although the private blockchain performs better than the public blockchain, however, a public blockchain should be utilized in a real-world scenario since it is truly decentralized.

   If nodes are permitted, they have the ability to read and write on the private blockchain. The role of the miners is to verify blockchain transactions and keep the blockchain secure by maintaining the blockchain database copies. Access control policies can be accessed by the nodes globally and can store these policies by using the interface of the blockchain. The data once stored in the blockchain is immutable and easily available to all the participants.

4. **Agent Node:** In our system, there is a special type of node called the agent node whose role is to implement the single smart contract in our proposed architecture. The agent node oversees executing the smart contract during the whole operation of the access control system. Once the smart contract is approved into the blockchain platform, the agent node obtains an address inside the blockchain that identifies the smart contract. To engage with a smart contract, every node on the blockchain network must know its address.

5. **Managers:** There exist some blockchain participants in the proposed system known as managers that manage the permission rights for a group of IoT devices. In contrast to mining nodes, managers, also known as light-weight nodes, do not validate blockchain transactions or retain transactional data. Consequently, in our system, the restrained devices can also act as managers despite the imitations of their hardware. Additionally, utilization of managers' hardware resources can be reduced by using our method since the managers do not need to be connected to the blockchain constantly.

   A manager can be any participant; nevertheless, the IoT devices which are registered must be under the control of the manager. Due to this, managers can be avoided from registering to devices which are under their control. Moreover, at least one registered manager in the system must have all the registered IoT devices under them. If that is not the case then the device will not be able to manage by any participant, and at the same time, many managers can oversee the same IoT device.

   The managers can define certain permissions for access control of the IoT devices once these devices are registered under the control of the manager.

6. **Smart Contract:** The system of access management is managed by the activities of the sole smart contract created in the system. All operations for the access management system are defined by smart contracts, which are activated by blockchain transactions. Once an action on the access control is initiated by a transaction, the miners will store the transaction data, which will be available worldwide. The smart contract's executing activities, as well as the smart contract itself, are both publicly available.

Furthermore, only managers, in our system, have the right to define new policies by interacting with the smart contract.

3.6 CHALLENGES OF THE INTEGRATED TECHNOLOGIES

The important challenges that arise and must be addressed while integrating IoT and blockchain technology will be discussed in this section. Integrating IoT with the blockchain technology is not straightforward. Blockchain technology was developed for an Internet scenario with high-performing machines, which is contradictory to the IoT technology. The devices must be able to operate with the cryptocurrency and must have the capability to digitally sign transactions which is a blockchain requirement. In this section, we highlight and outline the key issues that occur when merging blockchain with IoT.

1. **Scalability and Storage**

    While the scalability and the storage ability of blockchain is debatable, however, in the case of IoT applications, the low power and the constrained resources make these challenges much bigger. By looking at this scenario, it seems unsuitable to use blockchain for IoT applications; still the mentioned limitations can be completely avoided or assuaged by some methods. This limitation can be really challenging for IoT applications and for the integration with blockchain as the devices may produce a massive amount of information in real time. Therefore, these challenges must be addressed while integrating the two technologies [37].

2. **Privacy of Data and Anonymity**

    Most applications of IoT deal with sensitive data, for example, an application of heart rate monitoring; hence, it is important to address the issue of privacy and anonymity of the data. To address the issue of identity management, blockchain is considered to be the best solution for IoT; nevertheless, in the case of Bitcoin, some applications may require anonymity.

    In public blockchains, the issue of privacy of the data has been discussed already along with other solutions. IoT devices face a major challenge of data privacy since these devices start at the collection of data and then use that data for various applications. It becomes a challenge to store and secure the device, its data, and making sure that only authorized people have access to it is another challenge as security software of cryptography must be integrated into the device [38].

3. **Smart Contracts**

    Smart contracts are one of the most important uses of blockchain technology; however, there exists many challenges that must be addressed. IoT devices may benefit from smart contracts in a variety of ways, although integrating them with IoT applications is complicated [39,40].

    Furthermore, oracles are used while working with smart contracts; the role of these special entities is to establish trust and provide data in actual real-world scenarios. Due to the instability of the IoT devices, the validation of these smart contracts could be compromised. Additionally, these smart contracts could be easily overloaded if multiple data sources are accessed. Currently, smart contracts do not address a huge amount of computation and share resources to distribute tasks even though they are decentralized and distributed.

4. Consensus Algorithms

It is challenging to engage directly in consensus techniques such as PoW due to the restricted nature of IoT devices in various IoT applications. For consensus algorithms, there exists a wide range of proposals; however, they have not been tested and are immature. In a blockchain network, depending on which sort of consensus protocol is used, the resource requirements are determined. Normally, these tasks are allocated to devices which do not have computational limitations or gateways which have the ability to perform these functions. This functionality can be provided by off-chain methods which handle the data outside the blockchain to reduce the computational time in a blockchain network [41].

3.7   CONCLUSION

The IoT system has limited computational resources and generates a huge amount of data. Due to this, these devices consist of several security risks that must be addressed. In this book chapter, we identified and described the various security threats related to the IoT devices. We introduced blockchain technology which provides certain abilities that help to make the IoT devices more secure against various attacks. We proposed an architecture that involves various types of nodes and can be used to integrate the IoT devices and the blockchain technology. Furthermore, we identified and described various challenges that come when blockchain and IoT are integrated.

REFERENCES

[1] V. Roman and J. Ordieres-Mere, IoT Blockchain Technologies for Smart Sensors Based on Raspberry Pi, *Proc. IEEE 11th Int. Conf. Serv. Comput. Appl. SOCA 2018*, pp. 216–220, 2019, doi: 10.1109/SOCA.2018.00038.

[2] L. Wang, et al., Cloud Computing: A Perspective Study, *New Gener. Comput.*, 28, 2, pp. 137–146, 2010, doi: 10.1007/s00354-008-0081-5.

[3] D. Gangwani and P. Gangwani, Applications of Machine Learning and Artificial Intelligence in Intelligent Transportation System: A Review, in *Lecture Notes in Electrical Engineering*, Springer, 2021, pp. 203–216.

[4] P. Gangwani, J. Soni, H. Upadhyay, and S. Joshi, A Deep Learning Approach for Modeling of Geothermal Energy Prediction, *Int. J. Comput. Sci. Inf. Secur.*, 18, 1, pp. 62–65, 2020.

[5] B. P. Rimal, D. Pham Van, and M. Maier, Mobile-Edge Computing vs. Centralized Cloud computing in fiber-wireless access networks, in *2016 IEEE Conference on Computer Communications Workshops (INFOCOM WKSHPS)*, pp. 991–996, Apr. 2016, doi: 10.1109/INFOCOMW.2016.7562226.

[6] V. Parmar, H. A. Sanghvi, R. H. Patel and A. S. Pandya, A Comprehensive Study on Passwordless Authentication, *2022 International Conference on Sustainable Computing and Data Communication Systems (ICSCDS)*, 2022, pp. 1266–1275, doi: 10.1109/ICSCDS53736.2022.9760934.

[7] Statista Research Department. Internet of Things (IoT) Connected Devices Installed Base Worldwide from 2015 to 2025, Aug. 03, 2016.

[8] C. Kolias, G. Kambourakis, A. Stavrou, and J. Voas, DDoS in the IoT: Mirai and Other Botnets, *Computer*, 50, 7, pp. 80–84, 2017, doi: 10.1109/MC.2017.201.

[9] J. Gubbi, R. Buyya, S. Marusic, and M. Palaniswami, Internet of Things (IoT): A vision, architectural elements, and future directions, *Futur. Gener. Comput. Syst.*, 29, 7, pp. 1645–1660, 2013, doi: 10.1016/j.future.2013.01.010.

[10] S. Sicari, A. Rizzardi, C. Cappiello, D. Miorandi, and A. Coen-Porisini, Toward Data Governance in the Internet of Things, in *New Advances in the Internet of Things*, R. R. Yager and J. Pascual Espada, Eds. Cham: Springer International Publishing, 2018, pp. 59–74.

[11] F. Hawlitschek, B. Notheisen, and T. Teubner, The limits of trust-free systems: A literature review on blockchain technology and trust in the sharing economy, *Electron. Commer. Res. Appl.*, 29, pp. 50–63, May 2018, doi: 10.1016/j.elerap.2018.03.005.

[12] S. A. Al-Qaseemi, H. A. Almulhim, M. F. Almulhim, and S. R. Chaudhry, IoT architecture challenges and issues: Lack of standardization, *FTC 2016 – Proc. Futur. Technol. Conf.*, December, pp. 731–738, 2017, doi: 10.1109/FTC.2016.7821686.

[13] B. Kehoe, Integrating the supply chain, *Mater. Manag. Heal. Care.*, 15, 8, pp. 26–29, 2006, doi: 10.1108/eum0000000000329.

[14] D. Gangwani, Q. Liang, S. Wang, and X. Zhu, An Empirical Study of Deep Learning Frameworks for Melanoma Cancer Detection using Transfer Learning and Data Augmentation, in *2021 IEEE International Conference on Big Knowledge (ICBK)*, December 2021, pp. 38–45, doi: 10.1109/ICKG52313.2021.00015.

[15] A. Mukherjee, Physical-Layer Security in the Internet of Things: Sensing and Communication Confidentiality under Resource Constraints, *Proc. IEEE*, 103, 10, pp. 1747–1761, 2015, doi: 10.1109/JPROC.2015.2466548.

[16] M. Conti, R. Di Pietro, L. V. Mancini, and A. Mei, Emergent properties: Detection of the node-capture attack in mobile wireless sensor networks, *WiSec'08 Proc. 1st ACM Conf. Wirel. Netw. Secur.*, pp. 214–219, 2008, doi: 10.1145/1352533.1352568.

[17] L. Xie, Y. Mo, and B. Sinopoli, False Data Injection Attacks in Electricity Markets, in *2010 First IEEE International Conference on Smart Grid Communications*, 2010, pp. 226–231, doi: 10.1109/SMARTGRID.2010.5622048.

[18] D. Swathigavaishnave and R. Sarala, Detection of Malicious Code-Injection Attack Using Two Phase Analysis Technique, *Int. J. Comput. Appl.*, 45, 18, pp. 8–14, May 2012.

[19] M. Pirretti, S. Zhu, N. Vijaykrishnan, P. McDaniel, M. Kandemir, and R. Brooks, The sleep deprivation attack in sensor networks: Analysis and methods of defense, *Int. J. Distrib. Sens. Networks*, 2, 3, pp. 267–287, 2006, doi: 10.1080/15501320600642718.

[20] F.-X. Standaert, Introduction to Side-Channel Attacks, in *Secure Integrated Circuits and Systems*, I. M. R. Verbauwhede, Ed. Boston, MA: Springer US, 2010, pp. 27–42.

[21] B. Collier, R. Clayton, D. R. Thomas, and A. Hutchings, Booting the booters: Evaluating the effects of police interventions in the market for denial-of-service attacks, *Proc. ACM SIGCOMM Internet Meas. Conf. IMC*, pp. 50–64, 2019, doi: 10.1145/3355369.3355592.

[22] Y. Zou and G. Wang, Intercept Behavior Analysis of Industrial Wireless Sensor Networks in the Presence of Eavesdropping Attack, *IEEE Trans. Ind. Informatics*, 12, 2, pp. 780–787, 2016, doi: 10.1109/TII.2015.2399691.

[23] H. Liu, A new form of dos attack in a cloud and its avoidance mechanism, *Proc. ACM Conf. Comput. Commun. Secur.*, pp. 65–75, 2010, doi: 10.1145/1866835.1866849.

[24] S. Gupta, A. Singhal, and A. Kapoor, A literature survey on social engineering attacks: Phishing attack, *Proc. IEEE Int. Conf. Comput. Commun. Autom. ICCCA 2016*, pp. 537–540, 2017, doi: 10.1109/CCAA.2016.7813778.

[25] S. Andy, B. Rahardjo, and B. Hanindhito, Attack scenarios and security analysis of MQTT communication protocol in IoT system, *Int. Conf. Electr. Eng. Comput. Sci. Informatics*, 2017, September, pp. 19–21, 2017, doi: 10.1109/EECSI.2017.8239179.

[26] M. S. Islam, M. Kuzu, and M. Kantarcioglu, Access pattern disclosure on searchable encryption: Ramification, attack and mitigation, *Ndss*, 20, pp. 1–15, 2012.

[27] B. Kannhavong, H. Nakayama, Y. Nemoto, N. Kato, and A. Jamalipour, A survey of routing attacks in mobile Ad Hoc networks, *Wirel. Commun. IEEE*, 14, pp. 85–91, 2007, doi: 10.1109/MWC.2007.4396947.

[28] M. N. Ismail, A. Aborujilah, S. Musa, and A. Shahzad, Detecting flooding based DoS attack in cloud computing environment using covariance matrix approach, *Proc. 7th Int. Conf. Ubiquitous Inf. Manag. Commun. ICUIMC 2013*, pp. 4–9, 2013, doi: 10.1145/2448556.2448592.

[29] N. Sayegh, A. Chehab, I. H. Elhajj, and A. Kayssi, Internal security attacks on SCADA systems, *2013 3rd Int. Conf. Commun. Inf. Technol. ICCIT 2013*, pp. 22–27, 2013, doi: 10.1109/ICCITechnology.2013.6579516.

[30] A. Barua, H. Shahriar, and M. Zulkernine, Server side detection of content sniffing attacks, *Proc. Int. Symp. Softw. Reliab. Eng. ISSRE*, pp. 20–29, 2011, doi: 10.1109/ISSRE.2011.27.

[31] M. Nawir, A. Amir, N. Yaakob, and O. B. Lynn, Internet of Things (IoT): Taxonomy of security attacks, *2016 3rd Int. Conf. Electron. Des. ICED 2016*, pp. 321–326, 2017, doi: 10.1109/ICED.2016.7804660.

[32] M. Nofer, P. Gomber, O. Hinz, and D. Schiereck, Blockchain, *Bus. Inf. Syst. Eng.*, 59, 3, pp. 183–187, 2017, doi: 10.1007/s12599-017-0467-3.

[33] M. Cash and M. Bassiouni, Two-tier permission-ed and permission-less blockchain for secure data sharing, *Proc. -3rd IEEE Int. Conf. Smart Cloud, SmartCloud 2018*, pp. 138–144, 2018, doi: 10.1109/SmartCloud.2018.00031.

[34] H. Sukhwani, J. M. Martínez, X. Chang, K. S. Trivedi, and A. Rindos, Performance modeling of PBFT consensus process for permissioned blockchain network (hyperledger fabric), *Proc. IEEE Symp. Reliab. Distrib. Syst.*, 2017, pp. 253–255, 2017, doi: 10.1109/SRDS.2017.36.

[35] M. A. Ferrag, M. Derdour, M. Mukherjee, A. Derhab, L. Maglaras, and H. Janicke, Blockchain technologies for the internet of things: Research issues and challenges, *IEEE Internet Things J.*, 6, 2, pp. 2188–2204, 2019, doi: 10.1109/JIOT.2018.2882794.

[36] P. Gangwani, A. Perez-Pons, T. Bhardwaj, H. Upadhyay, S. Joshi, and L. Lagos, Securing Environmental IoT Data Using Masked Authentication Messaging Protocol in a DAG-Based Blockchain: IOTA Tangle, *Futur. Internet*, 13, 12, p. 312, December 2021, doi: 10.3390/fi13120312.

[37] P. Cui, U. Guin, A. Skjellum, and D. Umphress, Blockchain in IoT: Current Trends, Challenges, and Future Roadmap, *J. Hardw. Syst. Secur.*, 3, 4, pp. 338–364, 2019, doi: 10.1007/s41635-019-00079-5.

[38] S. Roy, M. Ashaduzzaman, M. Hassan, and A. R. Chowdhury, BlockChain for IoT security and management: Current prospects, challenges and future directions, *Proc. 2018 5th Int. Conf. Networking, Syst. Secur. NSysS 2018*, 2019, doi: 10.1109/NSysS.2018.8631365.

[39] A. Aniso, A. Chaudhary, and S. Sahana, A Review of Defense against Distributed DoS attacks based on Artificial Intelligence Approaches. In *2021 IEEE 6th International Conference on Computing, Communication and Automation (ICCCA)*, pp. 32–38, IEEE, 2021.

[40] M. Kumar Singh, K. Deep Mishra, S. Sahana An Intelligent Real Time Traffic Control Based on Vehicle Density, in *International Journal of Engineering Technology and Management Sciences*, 5, 3, 2021, doi: 10.46647/ijetms.2021.v05i03.004. ISSN: 2581-4621.

[41] P. K. Sharma, N. Kumar, and J. H. Park, Blockchain technology toward green IoT: Opportunities and challenges, *IEEE Netw.*, 34, 4, pp. 263–269, 2020, doi: 10.1109/MNET.001.1900526.3.

4 Artificial Intelligence in Cloud Computing

Dinesh Soni and Neetesh Kumar

IIT Roorkee

CONTENTS

4.1 INTRODUCTION

Cyber-physical system (CPS) is defined as "Systems that integrate the cyber-world with the physical-world" as per the National Institute of Standards and Technology [46]. Because CPS DataStream has a big amount of data, machine-learning (ML) techniques are used to analyze and interpret it. In the CPS, ML approaches are employed in three domains: smart grid (SG), transportation, and industry. Artificial neural networks (ANN) can forecast trends by performing regression on a time series stream. ANN predicts energy consumption to help clients regulate demand. Support vector machine (SVM) is utilized in fault detection, forecasting, clustering, feature design, and in application domains such as SG and transportation since it has a short response time. Random forest (RF) is used to detect faults in manufacturing machines or products, as well as spurious electricity records from sensors. In the power system, a decision tree classification technique is utilized for fault detection, energy demand prediction, and bus travel time.

This chapter is organized into three sections. First, Section 4.2 contains six subsections namely, CPS structure, CPS components, CPS characteristics, domains of CPS, CPS challenges, and CPS security. Second, Section 4.3 contains cloud computing technologies that discuss cloud computing systems, ML/deep learning (DL)

DOI: 10.1201/9781003248750-4

51

systems, packages, libraries, models, frameworks, devices on edge computing, and learning techniques in cloud networks. Third, Section 4.4 contains artificial intelligence (AI) technologies that discuss ML-based cyber-physical attack detection. The subsection discusses detection of malware using ML algorithms that describe CPS malware detection using ML.

4.2 CYBER-PHYSICAL SYSTEM

4.2.1 CPS Structure

The term CPS was proposed by the National Science Foundation in the United States in 2006. The President's Council of Advisors on Science and Technology acknowledged the CPS strategy as a critical and promising transition toward the future of the network and IT domain in its 2007 report. A holistic view of CPS is shown in Figure 4.1. The physical world activities are to be monitored [8]. Embedded devices of the next generation process the data and communicate with their distributed environment are termed as cyber systems. The communication and other middle components are referred to as interfaces. It connects cyber systems to the physical world by using actuators, Analog to Digital Converter (ADC)/ digital-to-analog converter (DAC) converters, interconnected sensors, etc. [6].

A sensor node performs the operation of sensing, actuation, computation (using processors and memory) and also contains communication modules, along with a battery. These sensor nodes can either transmit raw data to the data fusion nodes or process this data using their processing capabilities and relay the needed portion to another sensor node [11].

The physical layer box should be understood as one or more physical layers that are dispersed across multiple zones. Each physical layer has a collection of physical processes that can operate individually or asynchronously. Sensors that monitor the measurements and controllers that control these processes must be secured against

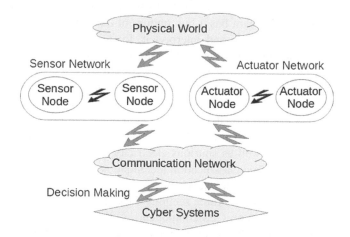

FIGURE 4.1 Holistic view of CPS.

intruders [20]. The output of a sensor is a determined measurement of its observation. The controllers that control the computations may be notified about the observed measurements. A controller must deliver the proper set of commands to an appropriate actuator for each observation received over the network layer. Actuators are located close to the physical processes with which they will interact for the sake of safety and security. Actuator attacks, external physical attacks, deception attacks, and Denial-of-Service (DoS) attacks along with traditional attacks are all major attacks on CPS [21]. Sensors and controls for managing physical dynamics are used in CPS. A communication infrastructure connects the dynamics.

4.2.2 CPS COMPONENTS

There are mainly four components in CPS that communicate with each other and work at two levels; cyber level and physical level [22]. Figure 4.2 shows the general structure of CPS. The components details are as follows:

1. **Physical Dynamics:** Physical dynamics alter with time, and the evolution law is established by the physical system, and the actions are taken to control it.
2. **Sensors:** The physical dynamics are sensed via sensors. Observations could be direct or indirect observations of the system state of physical dynamics.
3. **Controllers:** A controller receives sensor reports, computes, and then executes the control action.
4. **Communication Networks:** For the sensors to deliver their reports to the controllers, a communication network is required.

CPS is similar to a variety of research subjects and terminologies. Figure 4.3 depicts a number of concepts that are similar to CPS. Large datasets that are difficult to acquire, store, manage, and analyze using traditional procedures or database software tools are referred to as *big data* [27]. Billions of diverse networked smart devices, actuators, and detectors make a large-scale CPS. *Cloud computing* opens up

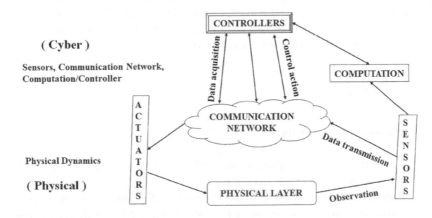

FIGURE 4.2 CPS structure.

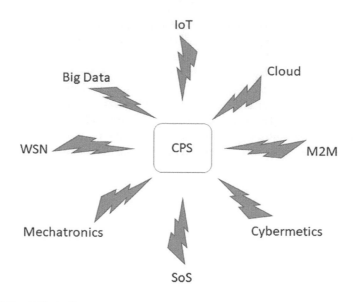

FIGURE 4.3 CPS similar concepts.

new possibilities for CPSs in terms of managing and operating averaged sensor data, and cloud-based decision-making approaches allow CPSs to improve system performance. Big diverse systems that are linked together for a shared goal and comprise autonomous elements that may be operated independently are known as *Systems of Systems* [24]. *Mechatronics* is a technique for designing, developing, and deploying composite engineering systems that merge information technologies into the physical environment. The study of processing and communication capabilities in both machinery and living things is known as *cybernetics*.

The Internet of Things (IoT) is concerned with observing things in the real world, using connectivity, and gathering data designed to handle objects which are not being handled effectively [30]. *M2M* refers to smart devices that communicate with one another across a network, such as PCs, smart sensors, mobile devices, embedded processors, and actuators. CPS integrates *wireless sensor networks* with intelligent road and autonomous vehicle navigation systems.

4.2.3 CPS CHARACTERISTICS

This section depicts an overview of characteristics of the CPS that distinguish it apart from other application domains.

1. **Reliability:** Software failures in the CPS can have disastrous implications. CPS is known for its high level of reliability.
2. **Security:** A security breach at a CPS involves data, but it can also have a significant impact on a person or property. Even if it is not directly accessible, CPS can be exploited.

3. **Real-Time Requirements:** In CPS, timing and deadlines are important. When specifying program correctness in the area of CPS, real-time requirements are a major differentiator [38].

4. **Data Representation:** The way data is represented has a huge impact on performance. A CPS interacts with hardware directly or communicates with other devices using well-defined protocols regularly. Data representation has a huge impact on performance, in addition to being a functional necessity.

5. **Constrained Environment:** CPS is known for working in resource-constrained situations. Memory is usually scarce, and CPU clock speeds are frequently slower than on comparable hardware platforms. When the energy source is a battery, power consumption offers a very different issue [46].

6. **Software Updates:** It's possible that updating the device may necessitate the use of specialized tools and techniques, as well as a significant financial investment. Downtime in an industrial control application or an Implantable Medical Devices (IMD) might have a powerful effect on CPS.

4.2.4 DOMAINS OF CPS

This section discusses key CPS applications as shown in Figure 4.4. *Smart manufacturing* is the application of embedded hardware and software to increase goods productivity or service delivery. Smart manufacturing is a very popular CPS application sector due to features such as international marketing, domestic, mass production, and economic growth [25]. The *emergency response* refers to managing public health risks, safety, and welfare, protecting the environment, vital infrastructure, and natural resources. Any civil or military aviation system, as well as its traffic

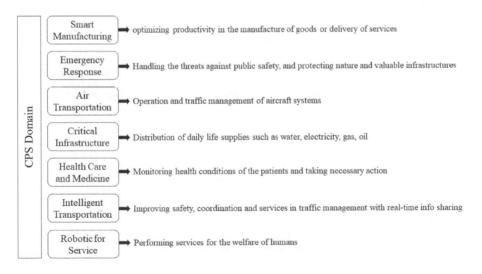

FIGURE 4.4 The functionality of each CPS domains.

management, is referred to as *air transportation*. The *critical infrastructure* refers to public facilities and characteristics required for the survival of the country. The most exciting application in the critical infrastructure framework is SG. The *health care and medicine* applications deal with a variety of components of a patient's physiology [46]. *Intelligent transportation* refers to enhanced sensor devices, message communication, processing, and control technologies used in the transportation domain. It improves coordination, services, and safety in traffic management by sharing actual data. The use of intelligent robotic systems to provide services for the well-being of individuals and appliances in autonomous or controlled remotely, excluding manufacturing tasks, is referred to as *robotic for service* [32]. The use of detectors, controllers, and decentralized control models to enable efficient remote monitoring and control of ventilation with security, and alarm systems in buildings is referred to as *building automation*.

4.2.5   CPS CHALLENGES

Figure 4.5 shows the relationship between the primary CPS challenges such as interoperability, predictability, reliability, sustainability, dependability, and security with their main characteristics. *Interoperability* means the module's ability to work collaboratively, transmit information, and use that information to execute certain services [8]. The degree to which a system's state, behavior, and functioning can be qualitatively or quantitatively predicted is referred to as *predictability*. *Reliability* tells the measure of correctness using which a system performs its task. Sustainability refers to the ability to endure without affecting the system's requirements, while also renewing and efficiently employing the system's resources. *Dependability* refers to a system's ability to accomplish required functions during operation without a major reduction in productivity or outcomes. The degree of trust placed in the entire system is reflected in dependability. *Security* means the ability of the system to provide

FIGURE 4.5   Future CPS challenges.

limited access to its resources and secure sensitive data from unauthorized access. *Maintainability* tells the ability of the system to be corrected in the event of a failure. *Availability* refers to a system's ability to be accessible even when failures occur. The ability of a system to perform well without causing any harm and risk to the system is referred to as *safety*.

Robustness refers to a system's ability to maintain its durable structure and survive in the presence of unexpected faults [12]. *Accuracy* gives the measurement of the match between the estimated and true outcome of the system. Compositionality is the property of a system's ability to be fully comprehended by looking at every aspect of it. Adaptability refers to a system's ability to vary its state to stay alive, by updating the configuration according to changing environmental conditions. Resilience means the ability of the system to continue operating and delivering services of acceptable quality in the face of internal or external challenges that do not surpass its endurance limit. Reconfigurability is the ability to change its settings in the event of a failure or response to internal or external requests [15]. The volume of resources like energy, cost, and time required by a system to perform defined functionality is referred to as efficiency. Integrity is the ability of the system to secure itself or the information it contains against illegal manipulation or change to maintain the information's correctness. The characteristic of permitting only authenticated users to access hidden data produced inside this system is known as *confidentiality*. *Composibility* is the attribute of distinct elements to be truly united, as well as their inter-relationships in the system [23]. Heterogeneity is a feature of a system that allows it to include a variety of interacting and interrelated components into a complex whole. Scalability means a system to continue to function normally despite a change in its size or increased workload and to reap the benefits of it.

Some other challenges are lack of standard, validation and verification tool, architectural design time management, system feedback, advanced query processing, software reliability, device interoperability, data extraction, security privacy, and reliability [27]. The issues that need further investigation are parameter variability, data workflow, practical implementation, etc.

4.2.6 CPS SECURITY

Attacks on CPSs can result in economic losses as well as casualties. As a result, examining and evaluating the most relevant publications to assess CPS security is essential [28]. The security framework for CPS technologies has three components like architecture layers, application scenarios, and measure, attack, defense, and control (MADC) types.

Figure 4.6 shows the CPS security framework that is explained below.

Architecture Layers: It contains three components perception, network, and application layers. The fundamental objective of the *perception layer* is to identify and measure objects before collecting and processing state information [39]. The *network layer* acquires the collected-data and delivers it to the application layer. It contains IoT protocol (ZigBee, Z-Wave, WiFi, Bluetooth, HDMI, etc.), lightweight crypto, traffic analysis methods, etc.

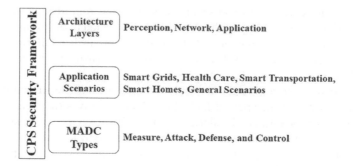

FIGURE 4.6 A unified CPS security framework.

The *application layer* receives information from this network layer and employs it to assist these applications. Authentication, access control, vulnerability analysis, and trusted computing are key parameters to study in this layer.

Application Scenarios: CPS technology has been widely used in a variety of scenarios like SGs, transportation, homes, and smart-industrial systems [34]. Embedded system devices contain various security features.

MADC types: There is a need to understand the entire offense and defensive processes to build an integrated prevention and control system.

- **Measure:** In CPS security, misconfiguration recognition, malware discovery, and risk assessments, device driver software, and integrated circuits are all critical safety measures.
- **Attack:** Researchers must also design new attack strategies to strengthen defense systems. Perception, application, and network layer attacks are the three types of attacks we categorize based on the location of the targets [23].
- **Defense:** Researchers have used various ways for detection and to block attacks in diverse structure layers [27].
- **Control:** Researchers have used several ways to distinguish structural layers. Device coupling, cloud service monitoring, trustworthy computing, key management, and encrypted communication schemes are all examples of control studies that have been conducted at the application layer.

4.3   CLOUD COMPUTING TECHNOLOGIES

This section summarizes the fundamentals of cloud properties, deployment models, services, characteristics, security issues, and challenges [1]. This helps to understand basic cloud computing systems. The cloud's properties are *user-centric* since the user owns the data in the cloud and can share it. Using *task-centric*, the user must concentrate on how the cloud will complete the task; it is powerful because many computational resources are connected, resulting in high computing power; it is accessible because many repositories are available in the cloud, so there is no single source of data; it is intelligent because the cloud has a large amount of data and retrieving tools, and it is programmable because the cloud has a large amount of data and many data mining tools are used to retrieve that data. The infrastructure-as-a-service (IaaS) contains many computation resources, memory needs, operating systems, and applications that users can control; software-as-service (SaaS) gives software utility services that can be used without installing on a PC and only requires an internet connection, and platform-as-a-service (PaaS) provides the ability to deploy applications and software that a user can control.

The deployment models present in the cloud are like: the private cloud is one in which a single user owns and operates the cloud, and the company uses it to maximize the use of computing resources and to work according to company needs; a *public cloud* is one in which third parties operate and own the cloud, and each user is assigned their computing capabilities, and a *hybrid cloud* can be created by combining public and private cloud. The specific community that shares common concerns and is shared by the organization is referred to as the community cloud. The characteristics of cloud such as *on-demand-self-service* offer the ability for users to provide computing resources as needed, *broad network access* provides cloud access using standards rules and can be accessed by various types of devices via the internet, *resource pooling* provides the ability to service many users using a multi-tenant model as needed, and *rapid elasticity* gives the features to scale up or down immediately in cloud. In *measured service*, the cloud monitors, controls, and reports resource usage, and it can also control and optimize resource utilization using metering and measuring capability. Based on some security issues, the user selects a cloud provider. Cloud security is privileged access to the user as the user information in the cloud has risks because of its data ownership and the user needs to be familiar with the regulatory systems; the cloud providers need to comply regularly with the security standards of a third party, while the user is accountable for the security solutions while choosing suppliers. The location of data should be known to the user, the data separation mechanisms should be used by the provider to separate encrypted data on one hard drive, and providers need recovery if a user data disaster occurs with certain recovery protocols. The details are shown in Table 4.1.

Now the next section discusses the edge computing domain's systems, packages, libraries, models, frameworks, and devices as shown in Table 4.2. The University of Toronto proposed CloudPath. It offers a variety of resources, as well as paths between devices and cloud data centers, which are mostly utilized for path computing [2]. It improves bandwidth consumption and reduces response time. PathExecute, PathStore, PathRoute, PathDeply, PathMonitor, and PathInit are among the modules found in the CloudPath node. PCloud integrates edge computing and storage resources to provide seamless support for mobile services. It increases the system's availability and optimizes resource configuration. The University of Wisconsin-Madison proposed ParaDrop [6], which provides effective resource use, multi-tenancy, and dynamic application management. It's utilized for IoT data analysis and offers privacy, low network latency, robust connections, minimal bandwidth, and location-aware services, among other things. SpanEdge allows programs to be run close to the data source, reducing network latency [2]. There are two sorts of communication: one between the worker and the manager (system management) and the

TABLE 4.1
Cloud Computing Systems

S. No.	Properties	Services	Deployment Models	Characteristics	Security Issues	Challenges
1	User-centric, task-centric, powerful, accessible, intelligent, programmable	IaaS, SaaS, PaaS	Private, public, hybrid, community	Shared-infrastructure, on-demand self-service, measured service, rapid elasticity, resource pooling, cost, reliable, sustainability, agility, maintenance, metering	Regulatory compliance, long-term viability, data-location, recovery, investigative support, privileged access, segregation	Information uses, privacy, accountability, connectivity, resource management, power consumption, reliability, energy-based resource scheduling, VM migration, workload prediction, network bandwidth, SLA violations, failure prediction, scheduling algorithms

TABLE 4.2

ML/DL Systems, Packages, Libraries, Models, Framework, Devices on Edge Computing

S. No.	Systems [2]	Packages [2]	Libraries [3]	Models [3]	Frameworks [3]	Devices [3]
1	EdgeX Foundry, CORD, Open VDAP, FocusStack, AirBox SpanEdge, LAVEA, Vigilia, VideoEdge, MUVR, CloudPath, PCloud, Cachier and Precog, ParaDrop, Firework	TensorRT, CoreML, PyTorch, MXNet	DL4J, SNPE, NCNN, Paddle-Mobile MACE, FANN	YOLO, DetectNet, MobileNet, GoogLeNet, DeepSense, AlexNet, LSTM, VGG, Deepface, SENNA, Faster R-CNN, VGG16, OpenALPR	TensorFlow Lite, Caffe2Go, MXNet, Core ML2, DeepThings, ML Kit, AI2GO, DeepIoT, DeepCham, SparseSep, DeepX, Edgent, daBNN, TensorFlow Federated, CONDENSA	NVIDIA GPU, SparkFun Edge, ARM ML, NVIDIA Jetson TX2, Raspberry-Pi kit, Google Coral Dev Board

other between the worker and the other worker (data transmission). CMU introduced the edge caching systems Cachier and Precog. System edge nodes are computable cache devices in the Cachier. It has three modules: an optimizer, a recognizer, and an offline analysis module. Cachier can determine the reduced cache size and modify the cache [2]. Precog is Cachier's extension.

A feature extraction module, an optimizer module, a profiler, and an offline analysis module are included in the Precog. AT&T laboratories presented FocusStack as a way to allow complex application deployment on IoT Edge devices. The Georgia Institute of Technology proposed AirBox. It offers a quick, flexible, safe, lightweight edge function loading service with privacy and security protection that is loaded on edge nodes. The MIST lab at Wayne State University developed fireworks. It offers a programming approach that includes both compute and management nodes. Azure IoT Edge enables cloud solutions to be migrated to IoT devices and offers hybrid cloud-edge analytics services. NVIDIA introduced TensorRT, a high-performance DL inference SDK that enables minimal latency and maximizes throughput in DL applications. Apple launched CoreML, which is used to incorporate a trained ML model into Apple products.

DL models, tree ensembles, generalized linear models, and SVMs are all supported by it [9]. It supports an iOS-based framework that is used to run ML on edge devices. Facebook offered PyTorch as an open-source ML platform for research purposes. It has GPU acceleration and deep neural network (DNN) for tensor processing. CMU and the University of Washington developed MXNet [2], which supports LSTM and convolutional neural network (CNN) networks. It is suitable for heterogeneous distributed systems because it is memory and computes efficient. This DL framework was adopted by Amazon. A lightweight solution as TensorFlow Lite is used for on-device inference for mobile devices and the edge [3] and runs on several CPUs/GPUs; it's ideal for distributed ML methods. Facebook introduced the Caffe2 DL framework. Caffe2Go is a DL model development framework built on top of Caffe2. Caffe2Go decreases the model size by using fewer convolutional layers. Google introduced the ML Kit framework as a mobile SDK framework. It's used for image labeling, text recognition, bar-code scanning, and smart response. The Xnor proposed AI2GO, which can execute DL models on low-resource devices and is utilized for on-device interfacing. DeepThings can execute CNN-based inference on edge devices and has a small memory footprint. DeepIoT is utilized to compress the DNN model. DeepX is a software accelerator that reduces resource overhead by creating unit blocks from the deep model network, which are then processed on mobile devices using heterogeneous processors. To improve DNN inference, Edgent is utilized to partition DNN computations amongst small mobile and edge devices and apply early exit at middle DNN layers. On ARM devices, daBNN is used to create a binary neural network. CONDENSA is a programmable system for compressing DL models, which uses Bayesian optimization to derive hyperparameters and can reduce memory footprint and execution time. Google Coral Dev Board was created to perform ML on the Edge TPU (co-processor) [3]. It has a 40-pin GPIO header on the baseboard for IoT devices, and the system-on-module runs Linux on a Cortex-A53 processor. Google, SparkFun, and Ambiq created SparkFun Edge to deliver real-time audio analysis, voice and gesture recognition at the edge [3]. As a visual processing unit, Intel Movidius is used in drones and intelligent security cameras. To construct ML, computer vision, image classification, and semantic segmentation, BeagleBone AI includes the Texas Instrument Deep Learning ML API. NVIDIA-Jetson–powered embedded computing boards and AI computing devices that are designed to process complicated real-time data and intelligent systems on edge devices [4]. The OpenMV Cam is a compact camera board with a low-power ARM Cortex-M7 processor for running machine vision algorithms [13].

There are many ML applications in the literature for this purpose as shown in Table 4.3. Cloud computing learning techniques play a major role. This section provides short details of learning techniques, challenges, parameters, and objectives from various research articles in the cloud computing field. ML affects many fields of research. ML techniques for improving cloud computing domains have been used in recent years [7]. Google suggested federated learning (FL) [5]. Multiple parties use FL techniques to develop an ML model and to maintain private data in collaboration. FL is multiparty, privacy-preserving ML. DL has a multilayer structure type, and this approach extracts accurate data from IoT-enabled devices that are placed in a complex networking environment and are used to increase performance on edge devices [6]. In cloud computing, ANN allows extensive operating capabilities and provides a quick and easy method of computing [8]. ANN is an ML technique that contains interconnected, multilevel neurons [19]. ANN training discovers many patterns and is used to detect intrusion in the cloud for cloud security. Cloud computing uses statistical ML with computer algorithms [10]. Statistical ML methods are used when a variable is statistically related and measured without a causal relationship [14]. Reinforcement learning (RL) works based on trial and error [17]. RL handles difficult real-world situations. It does not use statistical samples provided by an external guide, as happens in supervised learning. RL works as an agent-based approach and an agent works to maximize rewards based on own experience [20].

Parallel Q-learning maintains Q-matrix convergence, rewards, and strategies throughout training [18]. It is a technique of off-political RL. Q-matrix offers a strategy for state transition. Less training and computing complexity can easily be understood, easily modified to structure the storage strategy. SVM is best suited for unbalanced datasets and better algorithms for cloud intrusion detection [16]. Many sensors are used for data sensing in IoT. Many algorithms are applied to identify devices, which can extract device features.

4.4   AI TECHNOLOGIES

This section elaborates on various security designs issues and related ML techniques for CPSs applications [47]. It discusses various cyber-physical attacks at the software, physical, and network layers as shown in Table 4.4.

In [25], the author proposed RNN model sensor detection techniques where the intruder targets the vehicle. It explored several wheel speed sensor attack situations, with the proposed detection approach classifying the position and number of attacked sensors. In [26], the authors explored a malicious image injection bout in contrast to a visual-based autonomous driving system. In [27], the authors described a Bayesian-network–based attack-detection technique for water treatment systems, using data

TABLE 4.3

Learning Technique in Cloud Networks

Ref.	Technique	Challenges	Metrics/Parameters	Area
[4]	ML	Decreasing costs and floating revenue for operators, network slicing	Network delay, throughput, backhaul cost, power consumption	Edge caching
[5]	FL	Communication costs, resource allocation, privacy, security, dropped participants, unlabeled data, interference, asynchronous FL	–	Edge network
[6]	DL	Limited communication, non-IID training data, unbalanced contribution	Comm. rounds, scalability, accuracy, latency, comm. cost, computation, convergence rate, comm. load, fairness, latency, resource overhead, DL training, inference scalability, energy efficiency	Edge computing
[7]	ML	Heterogeneous backhaul/ fronthaul mang., infrastructure update, network slicing, standard datasets, transfer learning	Transmission power, data rate, interference, throughput, throughput, spectrum utilization, spectral efficiency, energy consumption, backhaul parameter, latency, cost	Wireless network
[8]	ANN	Wireless VR, unmanned aerial vehicles, and computing, IoT	Content correlation, caching efficiency, hit ratio, resources mang., path, channel, handover, content, computation, demand, LoS link	Wireless network
[9]	ML (nearest neighbor, naive Bayes, decision tree, SVM)	Parameter value optimization, data Prefetching, system performance	Temporal/spatial correlation, temporal/spatial sampling, freq. of instructions, performance tuning, prefetcher configuration, hardware events	Configuration for memory prefetchers

(Continued)

TABLE 4.3 (Continued)
Learning Technique in Cloud Networks

Ref.	Technique	Challenges	Metrics/Parameters	Area
[10]	Statistical ML (linear and LOESS regression)	Optimizing control parameters, model management, complex real-life workload/performance, internet data centers	Usage patterns, hardware failures, application changes sharing resources, queuing models, energy efficiency, G/G/1 queues, concurrency levels, and workload, requests, SLA threshold, workload, number of servers, performance, latency, gain controller	Optimal control for data centers
[11]	Supervised ML (modified PQR2)	Sudden changes in workloads management, discovery time ranges, time of day patterns, data skew	Query execution time, query plan, system load, span, query cost estimation, predict exec times of tasks and resource consumption	Predicting time and resources
[12]	ML (linear regression) + time series analysis	Resource autoscaling, real elasticity, cost-effectiveness in the pay-per-use, migration of web applications, queuing models, multi-tenant VMs	Web requests, resource demand, scaling, elasticity, effectiveness, M/M/m model, seasonal time patterns, web request distribution	Predict the number of requests
[13]	Error correction NN and linear regression	On-demand allocation, usage prediction, dynamic provisioning, trade-off b/w SLAs and constraints (VM setup overhead, cost-effectiveness)	PRED, R^2 Prediction, Max and Min EC2 instances	Predict resource usage patterns
[14]	Statistical ML (PRESS)	Large-scale cloud infrastructures, reduce resource waste	Fine-grained dynamic patterns, resource prediction accuracy, cyclic and non-cyclic workloads, load traces, over-under estimation errors, SLO violations rate, penalty functions	Predict resource demands

(Continued)

TABLE 4.3 (Coninued)
Learning Technique in Cloud Networks

Ref.	Technique	Challenges	Metrics/Parameters	Area
[15]	ML (SVM, NN, linear regression)	VM provisioning, public cloud infrastructure, database server	CPU utilization, response time, throughput, mean absolute percentage error, RMSE, percentage of observation	Resource demand prediction
[16]	ML (regression tree + boosting)	Database resource management for systems, optimal resource allocation	Service latency, response time, arrival rates	Adjust allocations and maximize profits
[17]	Reinforcement learning	VM Configurations, distributed RL	Memory-size, bandwidth, CPU-instances	Optimal configurations for VMs
[18]	Parallel Q-learning	Optimizing resource allocation, live virtualized test-bed	UTC, response time, arrival rates	Improving convergence times
[19]	ANN + SVM	Allocation and sizing of resources, storage I/O latency	MIMO queue model, CPU-memory limit, I/O latency	Performance and resources allocation
[20]	GA + Support vector regression	Internet, auxiliary memory, a resource optimization module	Response time, CPU, RAM	Reduce service response times

TABLE 4.4
ML-Based CPS Attack-Detection Methods

	Ref.	CPS Domain	Defense Against Concepts	ML-based Model	Used Validation
Physical layer	[25]	Automobiles	Sensory devices	RNN	Experimentation
	[26]	Automobiles	Sensory devices	CNN	Modeling
	[27]	Water	Sensory devices	Numerous (Bayesian Network)	Modeling
	[28]	HVAC	Sensor	SVM	Modeling
	[29]	Power	Sensor	DRE	Simulation
Network layer	[42]	Automobiles (VANET)	DoS	DNN	Simulation
	[30]	ICS	Numerous	Numerous (SVM)	Experimentation
	[44]	Pipeline of gas	Numerous	K-nearest neighbor	Actual data
	[45]	ICS	Numerous	LSTM-FNN	Experimentation
	[46]	Energy	Various	CNN	Experimentation
	[31]	Energy	DoS	SVM	Modeling
Application layer	[32]	–	Application programs	Numerous	Experimentation
	[35]	–	Application programs	CNN	Experimentation
	[37]	–	Application programs	Softmax	Experimentation
	[38]	–	Processing hardware	Numerous (Logistic Regression, SVM, MLP)	Experimentation
	[39]	–	Computing hardware	CNN	Experimentation

from many actuators and sensors. In [28], the authors proposed a combination of physical-model and learning-model detectors. On the learning-model–based detector, a one-class SVM model is applied. The authors [29] proposed a density ratio estimation (DRE)-based anomaly detection method for AC microgrid. The authors [42] proposed a hybrid jamming attack model. To inform the administration of threats, the suggested detector uses an SVM and k-means–based clustering algorithm, enabling a classification of distinct forms of network attacks [30]. In [44], the authors presented a two-stage packet-level anomaly model detector. To detect correlated network attacks, in [45], the authors proposed combining learning approaches using a forward neural network (FNN) model and an LSTM model. In [46], a CNN-based network attack detector for supervisory-control and data-acquisition (SCADA) networks was proposed. The authors [31] suggested single-class SVM to secure against DoS assaults on SG-SCADA systems. The authors [32] proposed a lightweight ML-based software anomalies analyzer for an embedded system in a camera application. The efficiency of several ML-based side-link detection techniques for guarding against mini-architectural side-link attacks was investigated in [33]. The authors [34] proposed an ML-based cache side-link attack-detection technique for realistic computational load situations. The authors [35] proposed a CNN model for detecting malware that classifies benign and malignant application software. The authors [36] suggested a machine-learning analyzer for malware detection in a computer by using a temperature sensor. The authors [37] proposed a softmax classification algorithm-based side-channel cache-attack recognition technique. To analyze CPS attacks in computer systems, the authors [38] proposed ML-based attack categorization techniques. The authors [39] proposed a CNN-based online row hammer detection that learns complex behavior in DRAM accesses by monitoring suspicious DRAM access patterns.

4.4.1 DETECTION OF MALWARE USING ML ALGORITHMS

The purpose is to use the SVM to detect anomalies in Android-based devices, so that trustworthy IoT services may be provided [40]. An approach for intrusion detection that uses behavior-based ML, and the use of a particular test bench for a CPS that simulates the identical features of a water system was demonstrated in order to collect data for the construction of a supervised ML model. An et al. [41] discusses the use of ML to identify malware in home routers. This proposes the detection of abnormalities based on multiple ML techniques to locate malicious traffic-utilizing communication data from industrial contexts as a dataset. A report released in 2018 provides a methodology for malware detection that is assisted by cloud environments [42]. The automated multilevel malware detection system (AMMDS) used memory forensic analysis techniques to monitor the early signs of malicious operation by recognizing unseen stealth activities in the host OS. The authors offer a lightweight cryptojacking classifier model based on the dendritic cell technique for detecting malware in CPS using dendritic cell algorithm (DCA). After becoming aware of the increasing trend in CPS to substitute PCs with mobile devices, the authors decided to develop a model for malware detection. The authors [43] concentrated on botnets detection, common destructive malware components, and the connection to the IoT. The detection system is built on a monitoring SVM-based system, which allows malware behavior patterns to be updated automatically [44]. In 2019, the author proposed research that aids in the prediction of malware threats in the CPS, particularly in the supply-chain area. The authors [45] focused on the use of ML in aerospace CPS applications to detect intrusions as shown in Table 4.5.

4.5 CONCLUSIONS

The CPS is a paradigm shift for present and future systems development that presents a significant influence on human communications with the actual world. This study presented relevant literature describing practical applications, structure, components, and characteristics, programing languages, advantages, domains, challenges, security, intrusion detection, SG, and dominating research topics in CPS. In addition, the mapping of CPS for the healthcare domain that will help in trends visualization, event prediction in healthcare, and to determine the differences between cloud-based

TABLE 4.5
CPS Malware Detection Using ML

Ref.	Environment	Classification Algorithm	Results	Future Work
[40]	IoT	SVM	In the IoT environment, SVM enhances the performance of many other classifier models in malware detection for Android-based devices.	–
	CPSs	SVM, NN, RF, J48, NB, BayesNet, LR	In water treatment contexts, the proposed method can detect the frequency of an attack as well as the type of attack with high accuracy.	Dataset expansion
[41]	IoT	Nineteen vector machine, principal components analysis, n-grams approach	With short fragments, SVMs and n-grams perform better. With huge fragments, all three algorithms are 100% effective, with no false positives.	Utilize smaller packages to expand the work.
	CPSs	SVM, RF, K-NN, K-means	Best detection results using SVM and RF	Dataset expansion
[42]	IoT	SVM	The application of SVM to improve the effectiveness of WMS	–
	CPSs	Automated multilevel malware detection method	AMMDS obtains 100% prediction performance, zero false positives, and a 5.8% CPS efficiency overhead	Dataset expansion
	CPSs	DCA	Cryptojackers in CPS can be identified and classified using a simple lightweight model	–
[43]	IoT	DNN	In IoT contexts, it improved supervised model performance and a semi-supervised approach for malware attack detection on Android-based devices using DL techniques	Dataset expansion (tagged and untagged data)
	IoT	DNN	The use of Tensorflow and DL (DNN) to detect botnets in IoT delivers a detection performance of above 97%	Increase the amount of current data in the datasets.
[44]	CPSs	SVM, independent component analysis, k-nearest, MEWMA, random forest	The combination of SVM with RF obtains prediction accuracy of up to 100%	–
	CPSs	SVM, decision tree	The use of decision trees with SVM algorithms to detect malware in the supply chain improves accuracy	–
[45]	CPSs	ML	Detection of attacks on CPS using historical attack data	Extend the availability of datasets

healthcare solutions with CPS-based healthcare solutions is also presented. The SG-CPS is a highly sophisticated structure that incorporates cyber and physical systems. The SG-CPS complexity requires the understanding of a comprehensive examination of its numerous domains. The most serious threat to CPSs is being infected with malware that turns them into botnets, numerous types of anomaly, and malware attacks in CPS. This chapter presented the importance of cloud computing technologies and current emerging ML algorithms in CPS environment domains. Because of increased data availability and the growth of significant computational power across devices, ML algorithms outperform classical algorithms due to better generalization ability. This chapter highlights the emphasis of ML techniques in the cloud computing domain. Future studies should focus on detecting botnets and the classification of malware in CPS using ML algorithms. There are still some significant challenges that need to be explored further to improve performance, cover new territory in the CPS domain, and cloud computing using AI.

REFERENCES

[1] Arockiam L, Monikandan S, and Parthasarathy G. Cloud computing: A survey. *International Journal of Internet Computing*, 1(2):26–33, 2011.

[2] Fang L, Guoming T, Youhuizi L, Zhiping C, Xingzhou Z, and Tongqing Z. A survey on edge computing systems and tools. *Proceedings of the IEEE*, 107(8):1537–1562, 2019.

[3] Murshed MG, Christopher M, Daqing H, Nazar K, Ganesh A, and Faraz H. Machine learning at the network edge: A survey. ACM Computing Surveys (CSUR), 54(8):1–37, 2021.

[4] Liying L, Guodong Z, and Rick S B. A survey of caching techniques in cellular networks: Research issues and challenges in content placement and delivery strategies. *IEEE Communications Surveys and Tutorials*, 20(3):1710–1732, 2018.

[5] Wei Yang Bryan L, Nguyen Cong L, Dinh Thai H, Yutao J, Ying-Chang L, Qiang Y, Dusit N, and Chunyan M. Federated learning in mobile edge networks: A comprehensive survey. *IEEE Communications Surveys & Tutorials*, 22(3):2031–2063, 2020.

[6] Yiwen H, Xiaofei W, Victor C M, Leung, Dusit N, Xueqiang Y, and Xu C. Convergence of edge computing and deep learning: A comprehensive survey. *arXiv*, 22(2):869–904, 2019.

[7] Yaohua S, Mugen P, Yangcheng Z, Yuzhe H, and Shiwen M. Application of machine learning in wireless networks: Key techniques and open issues. *arXiv*, 21(4):3072–3108, 2018.

[8] Mingzhe V, Ursula C, Walid S, Changchuan Y, and Merouane D. Artificial neural networks-based machine learning for wireless networks: A tutorial. *arXiv*, 21(4):3039–3071, 2017.

[9] Shih-wei L, Tzu-Han H, Donald N, Chinyen C, Chiaheng T, and Hucheng Z. Machine learning-based prefetch optimization for data center applications. In *Proceedings of the Conference on High Performance Computing Networking, Storage and Analysis*, 1–10, 2009.

[10] Peter B, Rean G, Charles A S, Armando F, Michael I J, and David A P. Statistical machine learning makes automatic control practical for internet datacenters. *HotCloud*, 9:12, 2009.

[11] Chetan G, Abhay M, and Umeshwar D. PQR: Predicting query execution times for autonomous workload management. In *2008 International Conference on Autonomic Computing*, 13–22. IEEE, 2008.

[12] Jing J, Jie L, Guangquan Z, and Guodong L Optimal cloud resource autoscaling for web applications. In *2013 13th IEEE/ACM International Symposium on Cluster, Cloud, and Grid Computing*, 58–65. IEEE, 2013.

[13] Sadeka I, Jacky K, Kevin L, and Anna L. Empirical prediction models for adaptive resource provisioning in the cloud. *Future Generation Computer Systems*, 28(1):155–162, 2012.

[14] Zhenhuan G, Xiaohui G, and John W. Press: Predictive elastic resource scaling for cloud systems. In *2010 International Conference on Network and Service Management*, 9–16. IEEE, 2010.

[15] Akindele A B, and Samuel A A. Predicting cloud resource provisioning using machine learning techniques. In *2013 26th IEEE Canadian Conference on Electrical and Computer Engineering (CCECE)*, 1–4. IEEE, 2013.

[16] Pengcheng X, Yun C, Shenghuo Z, Hyun Jin M, Calton P, and Hakan H. Intelligent management of virtualized resources for database systems in cloud environment. In *2011 IEEE 27th International Conference on Data Engineering*, 87–98. IEEE, 2011.

[17] Cheng-Zhong X, Jia R, and Xiangping B. URL: A unified reinforcement learning approach for autonomic cloud management. *Journal of Parallel and Distributed Computing*, 72(2):95–105, 2012.

[18] Enda B, Enda H, and Jim D. Applying reinforcement learning towards automating resource allocation and application scalability in the cloud. *Concurrency and Computation: Practice and Experience*, 25(12):1656–1674, 2013.

[19] Sajib K, Raju R, Ajay G, Ming Z, and Kaushik D. Modeling virtualized applications using machine learning techniques. In *Proceedings of the 8th ACM SIGPLAN/SIGOPS Conference on Virtual Execution Environments*, 3–14, 2012.

[20] Chenn-Jung H, Yu-Wu W, Chih-Tai G, Heng-Ming C, and Jui-Jiun J. Applications of machine learning to resource management in cloud computing. *International Journal of Modeling and Optimization*, 3(2):148, 2013.

[21] Gian Antonio S, Andrea S, Simone P, Sean M, and Alessandro B. Machine learning for predictive maintenance: A multiple classifier approach. *IEEE Transactions on Industrial Informatics*, 11(3):812–820, 2014.

[22] Eneko O, Enrique O, Asier M, Pedro L-G, Asier P, and Pablo G B. Decentralised intelligent transport system with distributed intelligence based on classification techniques. *IET Intelligent Transport Systems*, 10(10):674–682, 2016.

[23] Mitchell Y, Yong Q, Jing Z, Ying G, Branko G C, and Steven W S. Automatic bearing fault diagnosis using particle swarm clustering and hidden Markov model. *Engineering Applications of Artificial Intelligence*, 47:88–100, 2016.

[24] Allen H T, Wai-Ki C, and Ling-Yau C. Detection of machine failure: Hidden Markov model approach. *Computers & Industrial Engineering*, 57(2):608–619, 2009.

[25] Shin J, Baek Y, Lee J, and Lee S Cyber-physical attack detection and recovery based on RNN in automotive brake systems. *Applied Sciences*, 9(1):82, 2019.

[26] Cai F, and Koutsoukos, X. Real-time out-of-distribution detection in learning-enabled cyber-physical systems. In *2020 ACM/IEEE 11th International Conference on Cyber-Physical Systems (ICCPS)*, 174–183. IEEE, 2020.

[27] Lin Q, Adepu S, Verwer S, and Mathur A. TABOR: A graphical model-based approach for anomaly detection in industrial control systems. In *Proceedings of the 2018 on Asia Conference on Computer and Communications Security*, 525–536, 2019.

[28] Paridari K, O'Mahony N, Mady A E D, Chabukswar R, Boubekeur M, and Sandberg H. A framework for attack-resilient industrial control systems: Attack detection and controller reconfiguration. *Proceedings of the IEEE*, 106(1):113–128, 2017.

[29] Chakhchoukh Y, Liu S, Sugiyama M, and Ishii H. Statistical outlier detection for diagnosis of cyber attacks in power state estimation. In *2016 IEEE Power and Energy Society General Meeting (PESGM)*, 1–5. IEEE, 2016.

[30] Maglaras L A, Jiang, J, and Cruz T. Integrated OCSVM mechanism for intrusion detection in SCADA systems. *Electronics Letters*, 50(25):1935–1936, 2014.

[31] da Silva, E G, da Silva, A S, Wickboldt, J A, Smith, P, Granville, L Z, and Schaeffer-Filho, A. A one-class NIDS for SDN-based SCADA systems. In *2016 IEEE 40th Annual Computer Software and Applications Conference (COMPSAC)* (Vol. 1), 303–312. IEEE, 2016.

[32] Yoon M K, Mohan S, Choi J, Christodorescu M, and Sha L. Learning execution contexts from system call distribution for anomaly detection in smart embedded system. In *Proceedings of the Second International Conference on Internet-of-Things Design and Implementation*, 191–196, 2017.

[33] Wang H, Sayadi H, Sasan A, Rafatirad S, Mohsenin T, and Homayoun H. Comprehensive evaluation of machine learning countermeasures for detecting microarchitectural side-channel attacks. In *Proceedings of the 2020 on Great Lakes Symposium on VLSI*, 181–186, 2020.

[34] Mushtaq M, Akram A, Bhatti M K, Chaudhry M, Yousaf M, Farooq U, and Gogniat G. Machine learning for security: The case of side-channel attack detection at run-time. In *2018 25th IEEE International Conference on Electronics, Circuits and Systems (ICECS)*, 485–488). IEEE, 2018.

[35] Wang, W, Zhao, M, and Wang, J. Effective android malware detection with a hybrid model based on deep autoencoder and convolutional neural network. *Journal of Ambient Intelligence and Humanized Computing*, 10(8):3035–3043, 2019.

[36] Patel N K, Krishnamurthy P, Amrouch H, Henkel J, Shamouilian M, Karri R, and Khorrami F. Towards a new thermal monitoring based framework for embedded CPS device security. *IEEE Transactions on Dependable and Secure Computing*,19(1):524–536, 2022. doi: 10.1109/TDSC.2020.2973959.

[37] Cho J, Kim T, Kim S, Im M, Kim T, and Shin Y. Real-time detection for cache side channel attack using performance counter monitor. *Applied Sciences*, 10(3):984, 2020.

[38] Li C, and Gaudiot J L. Detecting malicious attacks exploiting hardware vulnerabilities using performance counters. In *2019 IEEE 43rd Annual Computer Software and Applications Conference (COMPSAC)* (Vol. 1), 588–597. IEEE, 2019.

[39] Chakraborty A, Alam M, and Mukhopadhyay D. Deep learning based diagnostics for rowhammer protection of DRAM chips. In *2019 IEEE 28th Asian Test Symposium (ATS)*, 86–865. IEEE, 2019.

[40] Ham H S, Kim H H, Kim M S, and Choi M J. Linear SVM-based android malware detection for reliable IoT services. *Journal of Applied Mathematics*, 2014, Article ID 594501, 10 pages, 2014. https://doi.org/10.1155/2014/594501.

[41] An N, Duff A, Naik G, Faloutsos M, Weber S, and Mancoridis, S. Behavioral anomaly detection of malware on home routers. In *2017 12th International Conference on Malicious and Unwanted Software (MALWARE)*, 47–54. IEEE, 2017.

[42] Zhou W, and Yu B. A cloud-assisted malware detection and suppression framework for wireless multimedia system in IoT based on dynamic differential game. *China Communications*, 15(2):209–223, 2018.

[43] Letteri I, Penna G D, and Gasperis, G D. Security in the internet of things: botnet detection in software-defined networks by deep learning techniques. *International Journal of High Performance Computing and Networking*, 15(3–4):170–182, 2019.

[44] Huda S, Abawajy J, Al-Rubaie B, Pan L, and Hassan M M. Automatic extraction and integration of behavioural indicators of malware for protection of cyber–physical networks. *Future Generation Computer Systems*, 101:1247–1258, 2019.

[45] Maleh Y. Machine learning techniques for IoT intrusions detection in aerospace cyber-physical systems. In Hassanien, A., Darwish, A., El-Askary, H. (eds) *Machine Learning and Data Mining in Aerospace Technology*, 205–232. Springer, Cham, 2020. https://doi.org/10.1007/978-3-030-20212-5_11.

[46] Griffor E R, Greer C, Wollman D A, and Burns M J. *Framework for Cyber-Physical Systems: Volume 1, Overview,* Framework for Cyber-Physical Systems: Volume 1, Overview, Special Publication (NIST SP), National Institute of Standards and Technology, Gaithersburg, MD, [online], https://doi.org/10.6028/NIST.SP.1500-201 (Accessed November 27, 2022), 2017.

[47] Sagar, R H, Tuiba A, Aastha S, Krishna Sai Raj G, Subrata S, and A K Sagar. Revolution of AI-enabled healthcare chat-bot system for patient assistance. In *Applications of Artificial Intelligence and Machine Learning,* 229–249. Springer, Singapore, 2021.

5 Current and Future Trends in an Intelligent Transportation System with Applications of AI

Atif Saeed, Abdul Basit Aftab,
Faraz Junejo, and Imran Amin
Shaheed Zulfikar Ali Bhutto Institute of
Science and Technology (SZABIST)

CONTENTS

5.1 INTRODUCTION

The past decade has seen a meteoric rise in the number of electronic devices, and especially devices that are able to communicate with each other. These devices are

DOI: 10.1201/9781003248750-5

often mobile and with the combination of different sensors and actuators that are able to fulfill any task, even remotely. The year 2008 was a head-turner in this department as the total number of connected devices exceeded the human population [1]. This trend has not seen any decline and has seen tremendous growth even till this very second. The ease in connectivity has enabled each and every device to be in touch with each other using the Internet. This has given rise to a new branch of connectivity naming IoT (Internet of Things). As the number of devices is seeing rapid growth, it only makes sense that the data that they are built to collect is also increasing.

5.1.1 INTERNET OF THINGS (IoT)

The building blocks in the IoT are 'things'. These things are basically any device that can either send or receive some data. These may include physical actuators, sensors or the underlying processors. These devices need to be connected to a common network for efficient and effective exchange of data through machine-to-machine (M2M) communication. This range of this communication can vary from short distances enabled using Bluetooth, Wi-Fi, ZigBee or for longer distances that can be accessed using GPRS, 3G, 4G, LTE and 5G [2]. The sheer quantity of connected devices present in our daily lives poses an opportunity to use all this data and interconnected devices to something more useful and productive. As these devices are produced in a massive scale, the production cost is brought to a bare minimum. Using M2M communication protocols, the devices are capable enough of not only collecting raw data but also using advanced analytics to produce logical deductions. As with any other industrial application, it is imperative to strike a perfect balance between its production cost, robustness and power use. IoT applications are at an all-time high, one of the areas where IoT can be applied is in smart cities. The IoT infrastructure can be seen in Figure 5.1. Under the branch of smart cities, numerous avenues are touched upon by IoT applicability, which is mentioned below [3]:

- **Smart Homes:** Under this category, daily-life household applications are included such as refrigerators, television sets and washing machines. The purpose of implementing them at this level is to create a network of

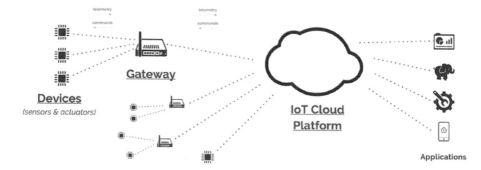

FIGURE 5.1   IoT infrastructure.

communicating devices to ensure automation and contextually aware decision-making for the utmost comfort of a user.

- **Healthcare Assistance:** The most important benefit of these IoT devices is their ability to collect data seamlessly. This is especially important for healthcare applications as patients' data regarding important health indicators can be collected using smartwatches and other sensors and be continuously monitored for anomalies.
- **Smart Transportation:** This industry has tremendous potential to be developed with IoT-based applications. Sensors embedded in routes or commuting vehicles can help recognize the shortest path, or the path with least traffic. This can even include vacant and occupied parking indicators for convenience of the general public. Weather and law and order situation can be major contributing factors for the traffic in third-world countries and can be updated using IoT devices.
- **Logistics and Supply Chain:** Industrial applications can be greatly enhanced with IoT-based applications. This can include using advances long- or short-ranged radio-frequency identification (RFID) devices to track parcels and products around every corner of the storage area and even to the production or dispersing unit.
- **Security and Surveillance Systems:** Continuous monitoring of sensitive places can be done by using imaging devices and images collected from these could be transmitted to control rooms or law-enforcing authorities for better surveillance.

5.1.2 MACHINE LEARNING (ML)

ML added a unique capability in traditionally running systems that helped industry work autonomously with more efficiency. This did not only make a system learn from previous experiences, but also induced decision-making capability, helping in a more productive and versatile environment. ML is nothing but a set of specially designed intelligent algorithms that enabled a system run and learn with every trial. The model most of the time is tested and trained on available datasets that help the system make decision through neuron just like the human brain. We can further seek it with an example that during early human life, when he touches a cup of hot beverage, his hand gets burned a bit; this event is learned by the early neurons present in human brain. Now till his last days, his brain is able to stop his fingers to touch some hot beverage as it might get burn; this is ML [4]. All the algorithms developed under the tree of ML are divided further into following streams.

- **Supervised Learning:** This is the kind of learning in which the model is fed with the possibilities and outcome both. The model is trained on both input and output so that on the basis of its relationship it could predict any floating input variable. The dataset required for supervised learning should include both, positive and negative images, most of the algorithm's search for features within the image such as threshold and the pixels to read and learn.

- **Unsupervised Learning:** This is a kind of learning in which the model is only fed with the possibilities; the dataset is so vast that the model learns on its own by running the finite number of iterations. To model such network is a very complex and difficult task and needs much more computational power.
- **Reinforcement Learning:** This learning style basically is based upon an incentive-based approach and on each step of the learning phase the algorithm is designed to increase the incentive. If the parameter and the output are aligned together then the learning phase is sped up for better results [5].

5.1.3 SENSOR TECHNOLOGY

Sensor technology has grown in popularity over the previous decade, attracting a lot of coverage. Sensors have been used in a variety of applications, including healthcare [6,7], agriculture [8], and forest [9,10], as well as automobile and marine monitoring

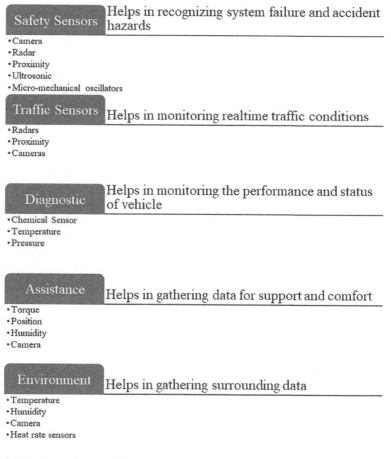

FIGURE 5.2 Classification of sensors.

[11,12]. Sensor technology aids the prototyping of a wide range of applications in transportation, including traffic control, safety, and entertainment. Numerous electrical devices under sensors and actuators categories have emerged that could be put to use in the transport industry. These include tire pressure sensor and rear-view cameras that could continuously monitor traffic and look for any mischievous activities [13]. These could eventually lead to better customer experience and citizens' trust on the overall city government and traffic reduction as well.

The sensors embedded in a vehicle cannot be limited to the only two listed in the above paragraph. More and more companies are looking for ways to make their vehicles 'smarter'. The current industry standard of including sensors is at about 60–100, but the way the need and demand for smart vehicles is increasing, it could certainly touch to around 200 [14].

5.1.3.1   In-Vehicle Sensors

Smart transportation system can only be built up once the correct identification of the problems involved with conventional traffic systems is done: (i) overcrowding and parking issues, (ii) extensive commuting times, (iii) harmful gases emissions and (iv) accidents are few of the areas of high interest in transportation systems. These are specifically important at not only individual levels but also at a collective level for the safety and security of everyone involved in the transportation network. Figure 5.3 illustrates few of the sensors installed in a typical vehicle.

5.1.4   IoT and ML in Making Our Current Transportation 'Smart'

The world is absolutely exploding with the potential and applications provided by IoT. IoT is not only solving existing problems but is also exploring up on new avenues for much better and smarter systems. The true prowess of IoT lies in the

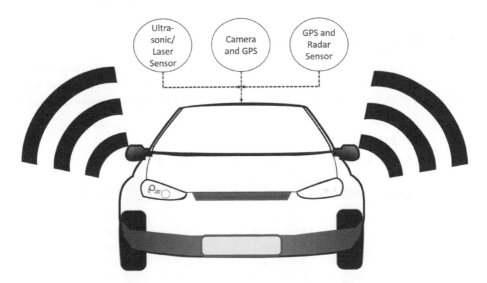

FIGURE 5.3   In-vehicle sensors.

fact that it is not restricted or limited to one or couple of systems but the fact that it has the ability to link every possible sensor and actuator to build up an actually complete system. This is the underlying concept behind the inception of IoT. Entire cities can get network-based connectivity. One area that could be greatly improved by this technological advancement is of the transportation system. The expanse of a typical and entire transportation system of a city is so tremendous and vast that it only makes it logical to implement IoT and reap the rewards of it leading to intelligent transportation systems (ITS). An extremely important area that could be worked upon by using IoT is the department of route optimization. This could be done by using the input from drivers' hand-held devices or with any other sensor along the commuting route for correctly analyzing the traffic situation for prompt update to future passengers on that route [15] and eventually leading to lesser fuel consumption and emission of harmful gases. This can be done by integrating sensors or cameras in street lights for effective monitoring and surveillance for security purposes as well [16]. Many experts have also proposed the idea of developing a parking system based on the principles of IoT as it can promptly notify the user in case of a vacant parking spot using simple IR and proximity sensors [16]. This could be of immense help and use to the general public and increase their productive time as they are not wasting it on a meaningless job just to park their car.

5.2   TRENDS OF SMART TRANSPORTATION SYSTEM

This section of the chapter provides a basic gist of the smart transportation system and all the limitations associated with it. As nearly as six different divisions of issues involved in it are identified, and their possible solutions are presented in this section. The solution is derived with the use of technologies such as ML and IoT algorithms. Table 5.1 depicts the categories of issues and their references along with it.

5.2.1   ROUTE OPTIMIZATION-NAVIGATION

Urban and metropolitan cities often have a major nuisance with their transportation systems called traffic congestion or overcrowding. This problem will only grow with the rapid increase in population and increased attention toward

TABLE 5.1

Smart Transportation Current State of the Art

S. No.	Smart Transportation System Application	Articles to Read
1	Parking	[16,18,19]
2	Road anomalies	[20–24]
3	Navigation	[25–29]
4	Infrastructure	[17,30]
5	Lights	[31–33]
6	Accident detection	[34–37]

migration to urban cities for finding livelihood. A solution that could be presented using IoT and ML algorithms is of route optimization. This is basically a technique that could provide any particular driver with the shortest route to any destination by accounting in all of the external factors. These factors can include time of day, weather conditions, road surface conditions and any other possible vehicle accidents. This could lead to ground-breaking development as it can prove valuable in not only providing comfort to the general public but also help in reducing road accidents. Emission of harmful gases can only be reduced if the vehicles are made to run for a shorter duration [15]. Several ways of implementing a route optimization algorithm have been proposed. One of them is by using an Markov decision process (MDP) algorithm that establishes a communication between vehicles for a group route. This will lead to entire groups of vehicles collaborating with each other with the same destination for a collective effort [38]. The other approach that has been proposed is a ML technique of implementing k-means and deep belief network (DBN) networks to streamline the traffic situation [25]. Several other literatures have provided background on different types of traffic forecasting. Short-term forecasting of traffic is handled in [39], and several other techniques such as FF-NN, Nonlinear Autoregressive (NARX) model and Seasonal Autoregressive Integrated Moving Average (BN-SARIMA} have been proposed for it. The same problem along with its counterpart are dealt in [28] with respect to application of four ML algorithms. These four algorithms are then tested among each other for performance and result evaluations. These four algorithms include baseline predictors, RF, FF-NN and a regression tree. FF-NN and baseline predictors are also used in [29] to estimate variations in time and total duration of the journey.

Researchers in [27] implemented a logistic regression model along with support vector machine algorithms and other models to predict traffic conditions. Lv et al. [40] used advanced data analytics and extraction to perform operations on the sensor data to predict the basic flow of traffic. The approach makes use of stacked autoencoder.

Swarm intelligence algorithms are a new department in computer science as they are nature-inspired algorithms and show great performance in modeling real-life systems. One of them is the ant colony optimization algorithm that could be used in conjunction with mobile crowd-sensing for ITS. The passengers are able to communicate with them for data regarding routes, surface conditions and other factors to find out less congested routes. This draws an analogy to how ants find their way to food by following the secretion of pheromone/chemicals [43].

Implementation of a different algorithm for route optimization is done in [42]. It makes use of a vehicular ad-hoc network that gathers traffic situation using mobile navigation devices and optimizes the route. This is similar in comparison to the above-mentioned techniques as it lets vehicles interact between them to share the traffic conditions.

The research in [43] takes a different approach of crowdsourcing the route planning by making use of a mobile application. An application by the name of Crowd Navi is installed on mobile devices, which wirelessly uploads the information about a particular traffic area to the Internet and among various vehicles.

5.2.2 PARKING

Another use of IoT in smart cities could be to provide a better parking infrastructure. This is a major issue in metropolitan cities where vehicle population is seeing rapid growth. Countless hours of the productive workforce of a city are wasted in trying to find a spot to park their cars. This could be done using IoT-enabled sensor and indicator devices that could update the information in a mainframe to distribute that information efficiently and effectively. Many experts also resort to imaging devices for detection of vehicles in a particular parking spot and then applying ML techniques to find the best possible vehicle for that position.

Niture et al. [44] provide a hardware-based idea to the concept of smart parking along with notification. A smart signboard is attached to the sensors and indicators setup that indicates the position of parking along with its status of being either vacant or occupied. Ultrasonic sensors are used to detect the presence of cars, and Wi-Fi technology is used to update the information in real time to a cloud server. Users are able to check the status of parking spots remotely which saves them a lot of time and hustle. The signboard is built around an LCD or LED display that has a microprocessor attached to it for it to translate the sensor data into actual signals for displaying it.

A similar approach is followed in [45] where ultrasonic sensors are used to check the presence of cars, and Arduino UNO is used as a microcontroller. A Wi-Fi module is used in tandem with the overall system to provide seamless communication protocol using the MQTT. The data is uploaded on the cloud server running on Thing Speak. Thing Speak is an IoT application that provides management as well as monitoring capability to users. Consumers can also be notified using an Android application on their mobile devices.

Performance evaluation of smart parking systems is done in [46], which validates the fact with a success rating of greater than 95%. The system is based upon the same ultrasonic sensors with use of ESP8266 controller with built-in Wi-Fi capabilities. REST API ensures a private cloud server to communicate the information of parking spots. User friendly application is developed for intuitive interaction between end users for maximum usability.

5.2.3 LIGHTS

Lighting is an essential part of any road to assist in driving at night times. Transportation vehicles are although fitted with lights, but it is extremely important to provide likewise facility to pedestrians and other smaller transport mechanisms. The concept discussed in this chapter is of smart street lights. They are supposed to run only when required to and can drastically save energy and fuel.

In the literature [33], an adaptable lighting system is put forward in which each lamp is Wi-Fi-enabled and is able to transfer a slew of data to the central control authority. The features that it proposes are smart turn on or off of the lighting by the varying amount of light intensity in the ambient conditions. This would enable energy and cost savings. The street lights can also be fitted with cameras for surveillance and security purposes.

5.2.4 ACCIDENT DETECTION AND PREVENTION

Accident prevention is major issue associated with the transportation system of any city. It is probably the most impacting event and hence must be taken into account very seriously as it directly affects the lives of people. A great deal of road accidents can be prevented if the drivers are focused on their commute, provided that they are abiding all the traffic rules. Smarter facilities in the transportation network can also help us detect already happened accidents at the earliest to supply them with emergency care measures.

The research article [35] uses a CHMM technique to exchange information between the vehicles in real time to prevent collisions and crashes. This way of preventing accidents but autonomously is presented in the research article [47] which also makes use of road sensors data fed into a FF-NN and regression tree algorithm. The basis of all techniques is the V2V (vehicle-to-vehicle) communication for an overall smarter decision [37]. Another important aspect is touched upon in [26] where along with V2V communication, blind-spot identification is also done using the Fully Convolutional Layer (FCN) method. This can also be implemented using the deep recurrent attention model (DRAM) method [48] which is based upon image processing and object detection. Driver consciousness [49] can also be monitored using image data and by using different computer vision techniques. These images can also be fed into a neural network to work up for most accurate and reliable results [36,50].

5.2.5 ROAD ANOMALIES DETECTION

Condition of road surface is another very important parameter that directly affects the flow of traffic and vehicle degradation. Effective and efficient identification of potholes and bumps will let users know beforehand of the caveats of any route and will present them with an option to change it accordingly.

The review paper [20] presents an elaborate take on the different deep learning techniques used in learning the surface of different pavements and roads. It makes use of as many as 12 of the most recent studies done in the same field. The neural networks that are employed are CNN and Deep convolutional neural network (DCNN). The techniques are pitted against each other with vital parameters such as performance, results, reliability and frameworks measured. The author finishes off with the opinion that CNNs perform the best when classifying pavement images.

5.2.6 INFRASTRUCTURE

With the recent advancement in IoT technology, conventional transportation systems can benefit greatly from it. The transportation system can be improved with much smarter working modes that can provide comfort at any level. The following paragraphs enlist the ways in which it can be done.

The research article [51] provides a mix of hardware and software to monitor bus fleets. This is done by using RFID tags, with each bus being uniquely identifiable and IR sensors fitted in the buses to count the number of passengers in a particular bus. GPS sensors are used to give information about the location of the vehicles. The data is collected and transmitted to a TI CC3200 microcontroller and eventually the cloud server. The controller module has a complimentary half fitted at each bus stop to display the arrival and departure of bus at the LCD at the bus stop. This can also be monitored remotely using the mobile application.

The IoT network can be combined with the social network to provide much better accessibility to the users. It is termed as SioV (Social Internet of Vehicles). The authors are of the opinion [17] that vehicular social network protocol can be used to reduce the amount of interference and crowdedness in an SioV network. This protocol can cover the MAC, physical and network layers to make the communication faster.

5.3   CASE STUDY: SMART TRANSPORTATION: THE CASE OF KARACHI

The findings of a comprehensive report published by United Nations states that the world population will increase to 5.2 billion in year 2050 from 2.6 billion in year 2010. As per this finding, one can clearly estimates that it is almost the double increase [53]. The basic necessities to live a quality life are directly linked with the resources government poses, and if the population increases with the current pace then there is no doubt that world is going to short fall on its resources, hence badly affecting the quality of life. The only solution out to tackle this global phenomenon is to develop such policies that enable the government to utilize their limited resources effectively, and henceforth, technology comes into play. The concept of smart and sustainable cities is to utilize the current advancement of mankind in engineering and extract the safe and higher standard of living for its residents [53].

Among the various problems that the urban areas are facing, one is traffic congestion. Traffic is a dynamic problem that can have various root causes: it can happen because of wrong/double parkings, heavy vehicle movement, some construction work, any roadside accident, etc. [54]. Currently, the authorities need to respond on time to ensure the flow of traffic. This gives a direction to researchers to work and find better solutions to this traffic congestion problem, and hence, the term smart transportation system came into being [55].

5.3.1   SMART TRANSPORTATION: CONCEPT

The system is a wireless sensor network (WSN) designed to sense different routine operations. Some of the salient features of such WSN is enlisted below [56].

- Loop detecting sensors can be used to sense the presence of vehicles, its volume, flow and speed.
- Radar sensor can be used on vehicles to detect possible obstacles and assure smooth traffic flow.
- IR sensors can be utilized to characterize automobile types through its echo measurement.
- Light detection and ranging sensors can be utilized within a vehicle to make the system autonomous and perform tasks such as obstacle detection and navigation without human interference.

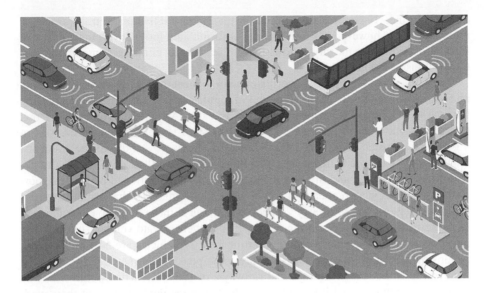

FIGURE 5.4   Smart transportation system.

With the current advancement in automobile industry, the vehicles nowadays are equipped with the sensors such as parking sensors, cruise control and lane change installed in them that could communicate with the advance transportation infrastructure as compared to the conventional that lacks this communication ability. These sensors provide more safety and degree of freedom. Moreover, some top-of-the-line vehicles have engine monitoring sensors that are directly connected to cloud and allow manufacturer to remotely monitor the engine performance. As Tesla already being a pioneer, it is predicted that the coming decade will be the era of unmanned vehicles comprising various in-built sensory networks that will help them in navigation, path planning, obstacle avoidance and optimized route selection, autonomously [57].

5.3.2   The Case of Karachi

The city of lights, Karachi, is the home to more than 15 million people as per the census conducted in 2017. It is the financial hub of the country, driving almost 70% of the total Pakistan's economy. Also, being the only port city till now, all of the country's supplies are transported from here. The big city is facing many big problems, and among them, the issue of traffic congestion is the highlighted one [58]. The city is facing various issues such as old transportation network, unwanted or illegal urban expansion and utilizing more than the roads capacity; all of these issues lead to current severe traffic problems.

Due to the uneven expansion of the city, the implementation of the smart transportation system will not be an easy task, but it has become a necessity [59]. There are routine traffic jams in the city causing its residents time, cost and affecting the overall environment. Moreover, with the 35% increase in the city population since 1947, the old transportation network is already working on its peak, and it's collapsing every other day. Since 1949, the Karachi's master plan inducted five major transportation-related projects; however, till date nothing is formulated [60]. Government's lack of interest in developing the city's infrastructure leads to its citizen relying on unmaintained transportation system.

The residents of the city rely on private sector to meet their daily mobility demand, and the lack of interest of the government in regulating this sector is also decreasing the overall standard of transportation day by day [61]. As per the optimized sharing, the total public transport shares 2%–3% of the total vehicles available on road at a given time, which could serve the demand of almost 40%–50% of the residents. However, the frustrated citizens are now more relying on their own vehicles rather than public transport, and this has disturbed the above-mentioned public transport sharing percentage a great deal. It can be understood in a way that a standard mini bus operating in a city can carry 27 people in it, but if those 27 people get annoyed with the facilities they are getting, they will use their private vehicles for mobility purposes; hence, in the place of a single mini bus, 27 vehicles are now on the road, collapsing the transportation infrastructure to its peak. The other way to divert the road traffic is the use of subways and railways; in 1964, Karachi Circular Railway project was initiated, and it helped its residents a great deal to move at their work places in a cheaper price. The project was functional till 1979, and after that again due to the lack of interest by the government in maintaining the railway infrastructure, regular delays in arrival and departure resulted in the decrease in its customers [62].

5.3.2.1 Proposed Framework for Karachi

The problem of traffic jam arises when the vehicles on a road surpass its capacity. Smart traffic system provides an efficient solution to this problem by not only alerting the residents about a possible traffic jam but also provides the next shortest route to the destination [63]. As we have already discussed in the previous section that the megacity of Karachi has seen enormous increase in its population and the lack of interest of the government, it is getting difficult day by day for the authorities to regulate traffic flow. Therefore, to give its citizens a sense of relief, the deployment of smart transportation system will be very beneficial.

With the influx of such great technologies, the transportation infrastructure is getting more and more advanced. This not only enabled the system to detect the dynamic behaviors, but also fast and accurate exchange of data. Moreover, with ML on board, the system starts predicting the high time of traffic jam of a particular road, hence alerting the users to take an alternate route. The proposed framework of smart transportation management system is shown in Figure 5.5.

The proposed model is designed keeping in mind the dynamic behavior of traffic flow in Karachi city. The proposed framework is divided in multiple administration sectors as per the landscape of the city. This administrative sector works as a subsystem and coordinates with the centralized information handling hub that processes the data received on regular intervals and conveys the information to its resident. This information can be conveyed in terms of multiple route or best time to travel. The junction will act as a brain of the system connected to various neurons (subsystems).

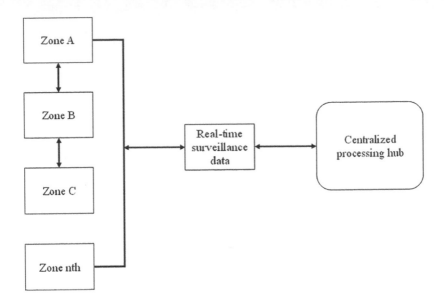

FIGURE 5.5   Proposed framework.

Multiple WSN are installed at various zones of the city in places like roundabouts, traffic signals and tall buildings and gather data and transmit it to subsystems using specific government-operated frequencies. This data can be stored at cloud for future reference and then sent to a central information handling station. These subsystems will work as the network layer, and the installed sensors at various points of the zone will act as the data layer.

Every pair of master and slave consumes a minimal time of microseconds to communicate and transmit data among the neighboring pairs. Figure 5.7 represents the average communication time between the neighboring pairs. With the increase in the number of vehicles in a designated area, the number of pairs also increases to transmit data effectively. Moreover, Figure 5.8 demonstrates the rate of data delivery with respect to the speed of vehicles. The graphs showed that more the distance between the two cars outskirts 500 m, the rapid the decrease in data transfer rate.

5.3.2.2 Concluding Remarks

The case study addressed the traffic congestion problem of one of the megacities of world, i.e., Karachi and provided a framework to eliminate its root causes with the integration of advanced technologies. The study highlights that the major reasons of traffic jams are uneven urbanization, massive population increase rate and lack of government policies to develop this fully utilized infrastructure. Hence, the ball is in the court of researchers to provide a technological solution to this problem, and hence, this framework is presented. This framework proposed the transformation of the already present transport infrastructure and making it smart by infusion of WSN. It is specifically designed with respect to the changing traffic dynamics of the city of Karachi and analyzed different way of communication to assure bidirectional flow of

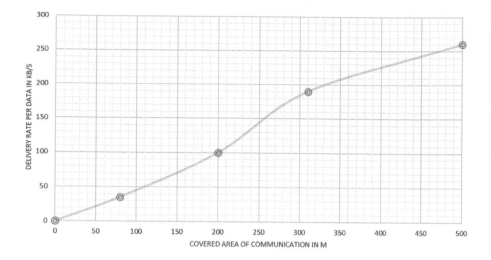

FIGURE 5.6 Time of delivery.

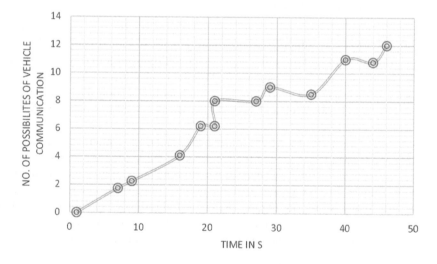

FIGURE 5.7 Average communication time between neighbors.

data. This will not only help the residents to get the real-time updates of traffic flow in city, but also help them to decide the alternate route in case of any congestion is sensed on a route.

5.4 CONCLUSION

The chapter sheds some light on the current scope of IoT in the transportation network and where its true potential lies. The chapter presents that IoT along with ML techniques could provide a ground-breaking development in the field of city's

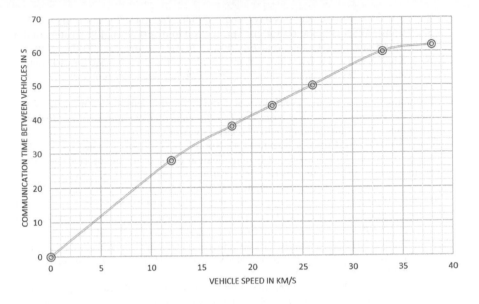

FIGURE 5.8 Rate of data delivery in given time.

infrastructure at a pretty wide scale. Applications such as smart lighting systems and parking applications can give rise to further applications.

REFERENCES

[1] M. Swan, "Sensor mania! The internet of things, wearable computing, objective metrics, and the quantified self 2.0," Journal of Sensor and Actuator Networks. 2012; 1(3):-217-253. https://doi.org/10.3390/jsan1030217

[2] Vangelista, L., Zanella, A., Zorzi, M. (2015). Long-Range IoT Technologies: The Dawn of LoRa™. In: Atanasovski, V., Leon-Garcia, A. (eds) *Future Access Enablers for Ubiquitous and Intelligent Infrastructures.* FABULOUS 2015. Lecture Notes of the Institute for Computer Sciences, Social Informatics and Telecommunications Engineering, vol 159. Springer, Cham. https://doi.org/10.1007/978-3-319-27072-2_7, pp- 51-58.

[3] Talari S, Shafie-khah M, Siano P, Loia V, Tommasetti A, Catalão JPS. A Review of Smart Cities Based on the Internet of Things Concept. *Energies.* 2017; 10(4):421. https://doi.org/10.3390/en10040421

[4] Mohammed, Mohssen, Muhammad Badruddin Khan, and Eihab Bashier Mohammed Bashier. *Machine Learning: Algorithms and Applications.* CRC Press, 2016.

[5] Badillo, Solveig, Balazs Banfai, Fabian Birzele, Iakov I. Davydov, Lucy Hutchinson, Tony Kam-Thong, Juliane Siebourg-Polster, Bernhard Steiert, and Jitao David Zhang. "An introduction to machine learning." *Clinical Pharmacology & Therapeutics* 107, no. 4 (2020): 871-885.

[6] Zhang, Yuan, Limin Sun, Houbing Song, and Xiaojun Cao. "Ubiquitous WSN for healthcare: Recent advances and future prospects." *IEEE Internet of Things Journal* 1, no. 4 (2014): 311-318.

[7] Alaiad, Ahmad, and Lina Zhou. "Patients' adoption of WSN-based smart home healthcare systems: an integrated model of facilitators and barriers." *IEEE Transactions on Professional Communication* 60, no. 1 (2017): 4-23.

[8] Bapat, Varsha, Prasad Kale, Vijaykumar Shinde, Neha Deshpande, and Arvind Shaligram. "WSN application for crop protection to divert animal intrusions in the agricultural land." *Computers and Electronics in Agriculture* 133 (2017): 88-96.

[9] A. A. Khamukhin and S. Bertoldo, "Spectral analysis of forest fire noise for early detection using wireless sensor networks," in *2016 International Siberian Conference on Control and Communications, SIBCON 2016 – Proceedings*, 2016.

[10] P. Bolourchi and S. Uysal, "Forest fire detection in wireless sensor network using fuzzy logic," in *Proceedings – 5th International Conference on Computational Intelligence, Communication Systems, and Networks, CICSyN 2013*, 2013.

[11] Guerrero, J. A., M. Cosío, A. Espinoza, E. Ruiz, J. D. Sánchez, J. Contreras, and J. I. Nieto. "GeoSoc: A Geocast-based communication protocol for monitoring of marine environments." *IEEE Latin America Transactions* 15, no. 2 (2017): 324-332.

[12] Pérez, Cristina Albaladejo, Fulgencio Soto Valles, Roque Torres Sánchez, Manuel Jiménez Buendía, Francisco López-Castejón, and Javier Gilabert Cervera. "Design and deployment of a wireless sensor network for the mar menor coastal observation system." *IEEE Journal of Oceanic Engineering* 42, no. 4 (2017): 966-976.

[13] USA Today, "NHTSA to require backup cameras on all vehicles," *USA Today*, 2014.

[14] Fleming, William J. "Overview of automotive sensors." *IEEE Sensors Journal* 1, no. 4 (2001): 296-308.

[15] A. Al-Dweik, R. Muresan, M. Mayhew, and M. Lieberman, "IoT-based multifunctional Scalable real-time Enhanced Road Side Unit for Intelligent Transportation Systems," in *Canadian Conference on Electrical and Computer Engineering*, 2017.

[16] Q. Wu, C. Huang, S. Y. Wang, W. C. Chiu, and T. Chen, "Robust parking space detection considering inter-space correlation," in *Proceedings of the 2007 IEEE International Conference on Multimedia and Expo, ICME 2007*, 2007.

[17] Jain, Bindiya, Gursewak Brar, Jyoteesh Malhotra, Shalli Rani, and Syed Hassan Ahmed. "A cross layer protocol for traffic management in Social Internet of Vehicles." *Future Generation computer systems* 82 (2018): 707-714.

[18] Amato, Giuseppe, Fabio Carrara, Fabrizio Falchi, Claudio Gennaro, Carlo Meghini, and Claudio Vairo. "Deep learning for decentralized parking lot occupancy detection." *Expert Systems with Applications* 72 (2017): 327-334.

[19] De Almeida, Paulo RL, Luiz S. Oliveira, Alceu S. Britto Jr, Eunelson J. Silva Jr, and Alessandro L. Koerich. "PKLot–A robust dataset for parking lot classification." *Expert Systems with Applications* 42, no. 11 (2015): 4937-4949.

[20] Gopalakrishnan, Kasthurirangan. "Deep learning in data-driven pavement image analysis and automated distress detection: A review." *Data* 3, no. 3 (2018): 28.

[21] Kulkarni, Aniket, Nitish Mhalgi, Sagar Gurnani, and Nupur Giri. "Pothole detection system using machine learning on Android." *International Journal of Emerging Technology and Advanced Engineering* 4, no. 7 (2014): 360-364.

[22] M. A. Al Mamun, J. A. Puspo, and A. K. Das, "An intelligent smartphone based approach using IoT for ensuring safe driving," in *ICECOS 2017- Proceeding of 2017 International Conference on Electrical Engineering and Computer Science: Sustaining the Cultural Heritage Toward the Smart Environment for Better Future*, 2017.

[23] M. Ghadge, D. Pandey, and D. Kalbande, "Machine learning approach for predicting bumps on road," in *Proceedings of the 2015 International Conference on Applied and Theoretical Computing and Communication Technology, iCATccT 2015*, 2016.

[24] Ng, J.R., Wong, J.S., Goh, V.T., Yap, W.J., Yap, T.T.V., Ng, H. (2019). Identification of Road Surface Conditions using IoT Sensors and Machine Learning. In: Alfred, R., Lim, Y., Ibrahim, A., Anthony, P. (eds) Computational Science and Technology. Lecture Notes in Electrical Engineering, vol 481. Springer, Singapore. https://doi.org/10.1007/978-981-13-2622-6_26

[25] Yang, Jiachen, Yurong Han, Yafang Wang, Bin Jiang, Zhihan Lv, and Houbing Song. "Optimization of real-time traffic network assignment based on IoT data using DBN and clustering model in smart city." *Future Generation Computer Systems* 108 (2020): 976-986.

[26] D. Kwon, S. Park, S. Baek, R. K. Malaiya, G. Yoon, and J. T. Ryu, "A study on development of the blind spot detection system for the IoT-based smart connected car," in *2018 IEEE International Conference on Consumer Electronics, ICCE 2018*, 2018.

[27] Devi, Suguna, and T. Neetha. "Machine Learning based traffic congestion prediction in a IoT based Smart City." *Int. Res. J. Eng. Technol* 4, no. 5 (2017): 3442-3445.

[28] Hou, Yi, Praveen Edara, and Carlos Sun. "Traffic flow forecasting for urban work zones." *IEEE transactions on intelligent transportation systems* 16, no. 4 (2014): 1761-1770.

[29] J. Yu, G. L. Chang, H. W. Ho, and Y. Liu, "Variation based online travel time prediction using clustered neural networks," in *IEEE Conference on Intelligent Transportation Systems, Proceedings, ITSC*, 2008.

[30] D. N. Chowdhury, N. Agarwal, A. B. Laha, and A. Mukherjee, "A vehicle-to-vehicle communication system using IoT approach," in *Proceedings of the 2nd International Conference on Electronics, Communication and Aerospace Technology, ICECA 2018*, 2018.

[31] Jia, Gangyong, Guangjie Han, Aohan Li, and Jiaxin Du. "SSL: Smart street lamp based on fog computing for smarter cities." *IEEE Transactions on Industrial Informatics* 14, no. 11 (2018): 4995-5004.

[32] J. Nausicaa, "Smart street lighting system using IoT and cloud computing," *Int. J. Res. Appl. Sci. Eng. Technol.*, 2021.

[33] A. K. Tripathy, A. K. Mishra, and T. K. Das, "Smart lighting: Intelligent and weather adaptive lighting in street lights using IOT," in *2017 International Conference on Intelligent Computing, Instrumentation and Control Technologies, ICICICT 2017*, 2018.

[34] Celesti, Antonio, Antonino Galletta, Lorenzo Carnevale, Maria Fazio, Aime Lay-Ekuakille, and Massimo Villari. "An IoT cloud system for traffic monitoring and vehicular accidents prevention based on mobile sensor data processing." *IEEE Sensors Journal* 18, no. 12 (2017): 4795-4802.

[35] W. Liu, S. W. Kim, K. Marczuk, and M. H. Ang, "Vehicle motion intention reasoning using cooperative perception on urban road," in *2014 17th IEEE International Conference on Intelligent Transportation Systems, ITSC 2014*, 2014.

[36] B. Ryder and F. Wortmann, "Autonomously detecting and classifying traffic accident hotspots," in *UbiComp/ISWC 2017- Adjunct Proceedings of the 2017 ACM International Joint Conference on Pervasive and Ubiquitous Computing and Proceedings of the 2017 ACM International Symposium on Wearable Computers*, 2017.

[37] N. Dogru and A. Subasi, "Traffic accident detection using random forest classifier," in *2018 15th Learning and Technology Conference, L and T 2018*, 2018.

[38] K. S. Sang, B. Zhou, P. Yang, and Z. Yang, "Study of group route optimization for IoT enabled urban transportation network," in *Proceedings -2017 IEEE International Conference on Internet of Things, IEEE Green Computing and Communications, IEEE Cyber, Physical and Social Computing, IEEE Smart Data, iThings-GreenCom-CPSCom-SmartData 2017*, 2018.

[39] G. Fusco, C. Colombaroni, L. Comelli, and N. Isaenko, "Short-term traffic predictions on large urban traffic networks: Applications of network-based machine learning models and dynamic traffic assignment models," in *2015 International Conference on Models and Technologies for Intelligent Transportation Systems, MT-ITS 2015*, 2015.

[40] Lv, Yisheng, Yanjie Duan, Wenwen Kang, Zhengxi Li, and Fei-Yue Wang. "Traffic flow prediction with big data: a deep learning approach." *IEEE Transactions on Intelligent Transportation Systems* 16, no. 2 (2014): 865-873.

[41] S. Distefano, G. Merlino, A. Puliafito, D. Cerotti, and R. Dautov, "Crowdsourcing and stigmergic approaches for (swarm) intelligent transportation systems," in *Lecture Notes in Computer Science (including subseries Lecture Notes in Artificial Intelligence and Lecture Notes in Bioinformatics), 2018.*

[42] Chang, Ing-Chau, Hung-Ta Tai, Feng-Han Yeh, Dung-Lin Hsieh, and Siao-Hui Chang. "A vanet-based a* route planning algorithm for travelling time-and energy-efficient gps navigation app." *International Journal of Distributed Sensor Networks* 9, no. 7 (2013): 794521.

[43] Fan, Xiaoyi, Jiangchuan Liu, Zhi Wang, Yong Jiang, and Xue Liu. "Crowdsourced road navigation: Concept, design, and implementation." *IEEE Communications Magazine* 55, no. 6 (2017): 126-128.

[44] D. V. Niture, V. Dhakane, P. Jawalkar, and A. Bamnote, "Smart transportation system using IOT," *Int. J. Eng. Adv. Technol.*, 2021.

[45] Elsonbaty, Amira, and Mahmoud Shams. "The smart parking management system." *arXiv preprint arXiv:2009.13443* (2020).

[46] A. Araújo, R. Kalebe, G. Girão, I. Filho, K. Gonçalves, and B. Neto, "Reliability analysis of an IoT-based smart parking application for smart cities," in *Proceedings -2017 IEEE International Conference on Big Data, Big Data 2017, 2017.*

[47] M. Ozbayoglu, G. Kucukayan, and E. Dogdu, "A real-time autonomous highway accident detection model based on big data processing and computational intelligence," in *Proceedings -2016 IEEE International Conference on Big Data, Big Data 2016, 2016.*

[48] J. L. Ba, V. Mnih, and K. Kavukcuoglu, "Multiple object recognition with visual attention," in *3rd International Conference on Learning Representations, ICLR 2015- Conference Track Proceedings, 2015.*

[49] Ghosh, Arnab, Tania Chatterjee, Sunny Samanta, Jayanta Aich, and Sandip Roy. "Distracted driving: A novel approach towards accident prevention." *Adv. Comput. Sci. Technol* 10, no. 8 (2017): 2693-2705.

[50] M. Munoz-Organero, R. Ruiz-Blaquez, and L. Sánchez-Fernández, "Automatic detection of traffic lights, street crossings and urban roundabouts combining outlier detection and deep learning classification techniques based on GPS traces while driving," *Comput. Environ. Urban Syst.*, 2018.

[51] S. Geetha and D. Cicilia, "IoT enabled intelligent bus transportation system," in *Proceedings of the 2nd International Conference on Communication and Electronics Systems, ICCES 2017, 2018.*

[52] United Nations, *World Population Prospects 2019: Highlights*, United Nations, Depart ment of Economic and Social Affairs, Population Division 2019.

[53] H. Chourabi et al., "Understanding smart cities: An integrative framework," in *Proceedings of the Annual Hawaii International Conference on System Sciences, 2012.*

[54] K. Berdica, "An introduction to road vulnerability: What has been done, is done and should be done," *Transp. Policy, 2002.*

[55] Guerrero-Ibáñez, Juan, Sherali Zeadally, and Juan Contreras-Castillo. "Sensor technologies for intelligent transportation systems." *Sensors* 18, no. 4 (2018): 1212.

[56] V. W. S. Tang, Y. Zheng, and J. Cao, "An intelligent car park management system based on wireless sensor networks," in *SPCA 2006: 2006 First International Symposium on Pervasive Computing and Applications, Proceedings, 2006.*

[57] Jin, Jiong, Jayavardhana Gubbi, Slaven Marusic, and Marimuthu Palaniswami. "An information framework for creating a smart city through internet of things." *IEEE Internet of Things Journal* 1, no. 2 (2014): 112-121.

[58] Mage, David, Guntis Ozolins, Peter Peterson, Anthony Webster, Rudi Orthofer, Veerle Vandeweerd, and Michael Gwynne. "Urban air pollution in megacities of the world." *Atmospheric environment* 30, no. 5 (1996): 681-686.

[59] Ahmed, Qureshi Intikhab, Huapu Lu, and Shi Ye. "Urban transportation and equity: A case study of Beijing and Karachi." *Transportation Research Part A: Policy and Practice* 42, no. 1 (2008): 125-139.

[60] Griffin, Keith B. "Financing development plans in Pakistan." *The Pakistan Development Review* 5, no. 4 (1965): 601-630.

[61] Sohail, Maunder, D. A. C. Maunder, and D. W. J. Miles. "Managing public transport in developing countries: Stakeholder perspectives in Dar es Salaam and Faisalabad." *International Journal of Transport Management* 2, no. 3-4 (2004): 149-160.

[62] A. Hasan, "Land, CBOs and the Karachi circular railway," *Environ. Urban.*, 2009.

[63] T. Litman and D. Burwell, "Issues in sustainable transportation," *Int. J. Glob. Environ. Issues*, 2006.

6 Intelligent 5G Networks and Augmented Virtual Reality in Smart Transportation

A M Anusha Bamini
Karunya Institute of Technology and Sciences

G R Gnana King
Sahrdaya College of Engineering and Technology

J H Jensha Haennah
St. Xavier's Catholic College of Engineering

CONTENTS

6.1 INTRODUCTION

Various networking topologies, mediums and methods supported in wired and wireless telecommunication are used to sustain information transmission between the distances among two or more nodes that form a network. The revolution in telecommunications, in specific wireless mobile communication, consumes to be substantial evolution as it has progressed over numerous generations, as mentioned in Figure 6.1.

DOI: 10.1201/9781003248750-6

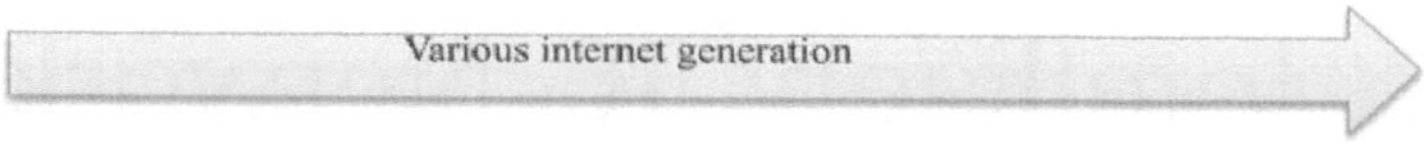

1G	2G	3G	4G	5G
1980's	1990's	2000's	2010	2018
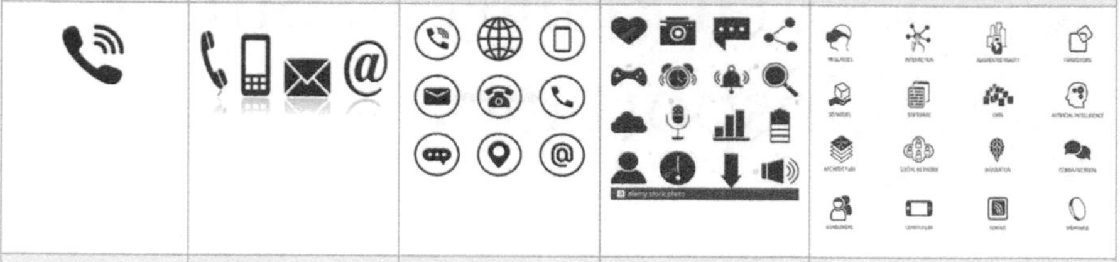				
Analog communication	Digital communication	Broadband	High speed data rates	Connected devices
2.4 Kbps	64 Kbps	3.1 Mbps	100 Mbps	10 Gbps

FIGURE 6.1 Internet generations.

On the basis of technical implementation, the evolutions are characterized for a particular standard, comprising novel methods and functions to distinguish it over to former generation [1].

Fifth generation (5G) was introduced in the year of 2018 [3,4]. It has been associated with plenty of new services, and a digital application provides a hyper-connected life to peoples. This technology can give an ultra-fast connectivity and creates a bond between peoples and industries. Due to the current pandemic situation, the need and growth of digital applications, collaborations and services increase dramatically. 5G mobile technology is working by overcoming the drawbacks in fourth generation (4G) technology. 5G and a combination of intelligent connectivity create smarter and better applications to make a life better. Globally around 50 countries such as China and Japan are expected to launch 5G mobile networks within 2025.

1G (first-generation) mobile network was hosted in the 1980s. Here analog technology was used for transmitting voice. Frequency division multiple access technology was supported. Signals were handled and transmitted by base stations. Voice call was the only service provided by 1G at that time. But 1G had various limitations. It did not support better data service and roaming facility. Maximum data speed provided by 1G was 2.4 kbps. Global roaming and poor voice quality were major drawbacks of 1G.

2G (second generation) introduced the concept of digital communication introduced in the 1990s. Analog-to-digital transmission increased the voice quality. Time division multiple access and collision division multiple access was supported. Quality of voice and data rate also increased in this generation. SMS, GSM, MMS services were introduced during this generation. With the support of WAP protocol, Internet access was provided to mobiles. SMS was also born with the onset of 2G mobile communication. The multimedia technology supported by 2G consumes high battery power.

3G (third generation) appeared in the 2000s. The very fast and wide Internet access was given by wireless data here. Technology supported was wideband code division multiple access. The big advantage of 3G was high data transmission rate. UMTS was proposed in 3G. While compared to GSM, UMTS offered two modes of bands like Frequency Division Duplex (FDD) and Time Division Duplex (TDD). Wide-frequency bands are supported here. Advanced multimedia applications are supported. All mobile phones are supporting VoIP and IP Multimedia Subsystem (IMS) systems. It provides the data transmission rate of 144 kbps.

4G was introduced in the year of 2010. This is mainly focused on Internet Protocol. It supports microwave access. 4G provides high-security, low-cost, high-speed Internet connection. Compared to other generations of network, 4G provides high band service. Data transmission speed is in megabytes and latency is in milliseconds. The Internet of Things (IoT) concept also introduced and implemented for smart home, smart city, smart office, etc. Due to the continuous growth in mobile communication technology, next-generation mobile network was introduced.

5G provides something new which was not previously available [2]. 5G has been introduced in the year of 2018. This completely creates new network architecture in combination with various technologies like artificial intelligence, IoT, multiple access edge computing and network function virtualization. More number of sensors is highly connected to create 5G network. Computing and data analytics techniques are introduced in 5G. Devices with 5G will be able to stay connected to the network anytime and anywhere, which opens up the possibility of connecting all devices on the network. Virtual and augmented reality concepts also live in 5G. For this reason, it is supposed that the basic 5G system strategy can support up to 1 million instantaneous connections per square kilometer. The objective in 5G network has providing 100 Mbps data for 97% users [5].

6.2   SMARTNESS OF 5G NETWORK

The 5G mobile network remains predictable for being the next big leap forward in mobile broadband. Based on research, it should overcome the drawbacks of 4G and LTE network. It is a high-speed, very low latency and high-bandwidth network. The network capacity of 5G is thousand times larger than 4G. Data transmission speed of 5G network was improved in two ways namely spectrum utilization and bandwidth expansion. From these two, bandwidth expansions have the better solution for improving network speed. The data rate of 5G can be increased by different metrics called edge rate, peak rate and aggregate data rate. This aggregate data rate represents the quantity of entire data transferred over the network. Data rate is improved thousand times in 5G network. Peak rate of 5G has between 10 and 20 Gbps. In addition to this, the latency of 5G is expected to be reduced to less than 1 ms. 5G is implemented to handle 100,000 connected people per square kilometer. At the maximum expected download speed of up to 20 Gbps, at a time, five users will be able to download entire movies in seconds. That much of network speed is provided by 5G. It can launch specified tasks and functions, such as remote control, precision medicine, connected cars, virtual and augmented reality and the IoT. It creates novel types of networks; these networks virtually connect with each other and with machines, electronic items, objects etc. 5G have basically three functions.

- To provide **Massive Internet-of-Things**, hypothetically comprising devices by ultra-low energy (more than 10 years life battery), ultra-low complexity (10 s of bps), and ultra-high density (per square km 1 million nodes).
- To provide **Mission-Critical Control**, hypothetically comprising ultra-high availability (greater than 99.999%), ultra-low latency (as low as 1 ms), and extreme mobility (up to 100 km/h).
- To provide **Enhanced Mobile Broadband**, hypothetically comprising great data rates (multi-Gbps peak, 100+ Mbps sustained) and higher capacity (10 Tbps of aggregate throughput per square kilometer).

6.3   5G INTELLIGENT TRANSPORTATION

Complexity and quantity of traffic is increased nowadays. To optimize the entire transportation, intelligent traffic is a best solution. 5G networks as well as artificial intelligence systems will interconnect the vehicles location, bicycles and people in real time, and reduce the possibility of accidents or collisions. Data such as weather, surface conditions, construction sites or traffic jams are transmitted in real time through the driver's helmet, and joined with artificial intelligence to support road users find better routes. 5G promotes the social benefits of connected vehicles, automation, sharing and electric vehicle, and can contribute to and truly become the core of the future development of autonomous vehicles. A survey [6,7] shows that in the year of 2035, 5G technology will provide 2.4 trillion dollars as the economic output. In Europe, the automotive transport will yield 220.5 billion euro.

The key concepts of intelligent transportation are as follows:

- Vehicular communication
- Automated driving
- Intelligent navigation
- Information society on road
- Isolated traffic light control

6.3.1   VEHICULAR COMMUNICATION

In recent years, all modern vehicles are working with the support of sensor platform, which extract the data from environment. This extracted information is handled by a computer system and then used for navigation, traffic management, pollution control and more. However, to attain quick processing, a very potent on-board computer is needed. The costly usage of equipment has been reduced, via Internet that ought to be potential for transferring data over to the cloud for accomplishing substantial processing data. Therefore, IoT can helpful to gather further information of traffic controlling centers, accompanying the data previously gathered from vehicles. Because of this reason, movement cloud computing worldview might be a tough spot to actually look at forthcoming 5G capacities [9]. Vehicles may trade data with

FIGURE 6.2   Subfields of vehicular communication.

various vehicles (represented as Vehicle to Vehicle (V2V) correspondence), due to the side of a road foundation, through the Internet, with a passenger and inside the equivalent manner with any component at intervals a sensible town. The term vehicle to everything or V2X is employed to sit down with of these kinds of conveyance communication, represented in Figure 6.2. By considering these methods, the most developments in vehicular communications are carried out, familiarizing appropriate circumstances for protection, quality and comfort.

6.3.2   Automated Driving

Before moving into an autonomous vehicle, there are several intermediate levels to interact with the vehicle [8]. Totally six stages of automated driving was recognized via Society of Automotive Engineers. These levels are converting no-automation vehicle into full-automation vehicle [13].

Briefly,

Level 0: Non-Automation – Driver continuously in control.

Level 1: Driver Support – Slight driving job accomplished by system.

Level 2: Fractional Automation – Driving tasks is monitored by driver dynamically.

Level 3: Conditional Automation – Only a small amount of control is given to driver rather than driving.

Level 4: High Automation – The driver is not necessary in this case.

Level 5: Full Automation – This is a completely automated system; driver is not required.

In principle, for levels 1 and 2 without V2X, communication-independent driving is possible, where only the human driver monitoring the driving environment. Level

5 is working rather than the need of wireless system. But for automatic driving, only an on-board processing and sensor are not enough. If there is no communication between vehicles then the uncertainty must be taken into account, because it is impossible to guarantee how other vehicles move or if pedestrians will cross the road next in few seconds; if information exchanging among automobiles is available, vehicles can use this information to diminish uncertainty. So independent drivers can benefit from local V2V connectivity and respond more quickly to maneuvers, avoiding collisions.

6.3.3 INTELLIGENT NAVIGATION

In autonomous vehicles, navigation guidance is provided by digital maps and geo-positioning system to drivers. According to this online traffic information, the efficiency of driving was improved to choose appropriate routes. This traffic information is calculated from data delivered by vehicles in traffic management or data captured by camera. Based on the support of 5G, IoT and big data, all traffic-related information is collected. Notifications will be received on the driver's mobile related to parking, restaurants, tourist spot, petrol bulk, ATM, hospitals, etc.

6.3.4 INFORMATION SOCIETY ON ROAD SAFETY

The upcoming products of independent vehicles are increased a lot in electro-mobility and automotive industry. However, some potential applications do not require a high degree of automation. Information delivered by on-board sensors by the means of cameras or lasers via a cloud connection also allows the driver to be informed by the alert app. For safe, efficient as well as comfortable driving, the local V2V and V2X communication are expected to be used by vehicles. These vehicles can easily recognize an accidental situation previously. With the support of V2X communication, all possible information's are shared to nearby vehicles. This possible information helps the driver to get knowledge about nearby vehicles on road. For example, one vehicle is moving on the road behind another vehicle. Suddenly one pedestrian may cross the road in front of the first vehicle. At that time, the first vehicle's camera detects pedestrian and shares that picture to the second vehicle to alert the situation. This can be done with the help of augmented reality. This practical implementation involves a high visibility, reliability, data rate and low latency. This augmented reality can also support for warning road hazard, warning diversion, warning road work, lane change and so on. Using Vehicle Roadside Unit (VRU) and vehicular cloud computing, accidents can be avoided.

6.3.5 ISOLATED TRAFFIC LIGHT CONTROL

In isolated control, traffic moves in a one-way path and does not communicate with the neighboring light signal [11,12]. The aim of isolated traffic was to minimize vehicle waiting time and maximize the number of vehicle traveled. Green signal period of traffic is represented as phasing to move vehicles from one direction to another. Signal phasing needs traffic direction to determine a vehicle's direction. Traffic light controller is modeled with sensors and traffic detectors. Traffic light control is connected with a number of parameters called direction, cycle length, phase length and offset [10].

Direction: Movement of vehicles from one direction to another direction.
Cycle Length: Amount of time needed to finish one signal phase.
Phase Length: Ratio of cycle length allotted to a signal phase.
Offset: Calculated a time difference between particular point and reference point in traffic.

Three types of traffic-isolated controls are represented as semi-actuated controller, fully actuated controller and volume-density controller.

Semi-Actuated Controller: Based on the intersection of minor and major streets, the density of traffic volume differs. To detect traffic in minor streets, a detector is necessary. A green light can be requested for minor streets and also it extends the green light till the vehicles are present in the street.
Fully Actuated Controller: This can be supported in a combination of two similar volumes of streets. Here a sensor is necessary to turn on the green signal. If there is no vehicle in the night time then the same signal will be continued or signal will be changed based on previously defined cycle time.
Volume-Density Controller: This is an advanced controller technique but similar to fully actuated controller. The volume of vehicles in green level as well as waiting vehicles' density in red phase is also measured. The controller can shorten the green stretching time or lengthen least green time by evaluating oncoming traffic. In order to respond correctly on traffic condition, sensors or detectors must be correctly and sufficiently positioned in front of the stop line.

6.4   IN-VEHICLE AND WIDE AREA CONNECTIVITY

The connectivity in a vehicle visualizes and enables multipurpose device connectivity at the same time. Each vehicle is occupied with the internal connectivity that is interconnected with actuator, sensor, radar, lidar, camera and some other devices. This intelligent 5G network transmits data between each devices and vehicles with very low latency. Also it is connected to the drivers', conductors' and passengers' mobile phones. Figure 6.3 shows the network connectivity of vehicles and road-side equipment. However, this wide area 5G network efficiently shares the spectrum resources.

6.5   SIGNALIZED ROAD NETWORK IMPROVEMENT

In fact, it is less likely that there will be a designated intersection, but rather there will be a network of roads with many intersections. Based upon the speed of vehicle and distance between intersections, a green light will be maintained at each junction for a good flow of traffic to avoid waiting of queuing vehicles. An algorithm called arterial

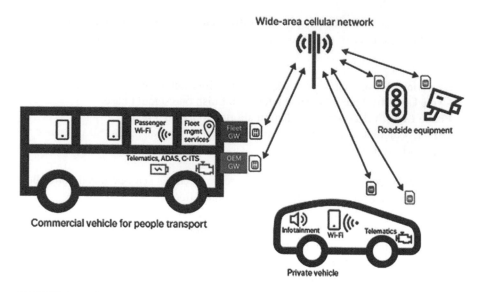

FIGURE 6.3   Connectivity for vehicle and network.

traffic light algorithm (ATL) is reducing traffic jams with the help of isolated traffic light controlling. In ATL, traffic flow is controlled by Intel-ligent Traffic Lights (ITL). It receives scheduling messages about the flow of traffic, vehicles coming to platform, starting and ending crossing time to ITL. The below-mentioned algorithm is used in traffic light. Based on the priority of arterial flow, throughput of traffic fluency increases. Equation (6.1) represents arterial factor (A_i) and saturation factor (S_i) to be combined to create combined factor (C_i). $\propto$ represents importance of these factors.

$$C_i = \propto \times A_i + \left(1 - \propto \times (S_i)\right) \tag{6.1}$$

Algorithm: Arterial Traffic Light Controlling Algorithm (AT$_{RL}$)

   Statistics used;

   T_{RL}: traffic light

   RM: reporting message

   N: no. of vehicles crossed prior T_{RL}

   S_{time}: starting cross-time for prior T_{RL}

   E_{time}: finishing cross-time for prior T_{RL}

   MF: maximum flow

   A_{time}: expected time of arrival

   L_{time}: expected time of leaving

LCFT: last cycle finish time

NCFT: next cycle finish time

Initial cycle phases of T_{RL} are programed using this algorithm:

Send RM of MF to following T_{RL} in a street;

if T_{RL1} accepts RM from T_{RL2}, it finds;

$A_{time} = S_{time} + distance\ (T_{RL1}, T_{RL2})/S_{time};$

$L_{time} = E_{time} + distance\ (T_{RL1}, T_{RL2})/S_{time};$

if $A_{time} < NCF\ T$

while Current Time $< A_{time}$ do

for planning upcoming cycle phases of T_{RL} and don't think MF;

end

use Equation 1 to compute Cf_i of every competing flow i;

while Cf_i of several traffic flows at $T_{RL} > 0$ do

let j is a traffic flow through maximum combined factor (Cf_i);

let i_1 and i_2 are the traffic flows which pass the joining with traffic flow j;

if $Cf_i 1 > Cf_i 2$ then

$P_{ji} 1 == ART\text{-schedule}\ (j, i);$

$Cf_i 1 = 0;\ t_i 1 = 0;$

else

$P_{ji} 2 == ART\text{-schedule}\ (j, i2);$

$C_{fi} 2 = 0;\ t_i 2 = 0;$

end

end

else

$C_{fj} = 0; t_j = 0;$

use the steps to plan the subsequent cycle phases of T_{RL};

end

6.6 ISSUES AND CHALLENGES

5G network replaces the drawbacks of 4G and LTE networks. D2D communication, millimeter waves and spectrum sharing were introduced in 5G.

- Storage as well as caching capacity of mobile devices can be improved.
- Network delay between devices can be decreased.
- Computational capability of mobile device and 5G network will be developed more.
- VR technology is applied to improve vision augmentation.

6.7 CONCLUSION AND FUTURE SCOPE

Development of new technologies and need of human play a vital role in 5G network, which supports and creates sophisticated techniques in smart driving. Efficiency of traffic is improved by traffic light optimization. Various techniques mentioned in 5G intelligent transportation enumerate road safety in smart city. The in-vehicle wide area connectivity creates a connection between vehicles, devices inside the vehicle, vehicle and traffic light and so on. Arterial traffic light controlling algorithm also controls the entire traffic network.

REFERENCES

[1] Prinima, D., and Pruthi, D.J. Evolution of mobile communication network: From 1G to 5G. *Int. J. Innov. Res. Comput. Commun. Eng.* 2016, 4, 224–227.

[2] Gupta, A., and Jha, R.K. A survey of 5G network: Architecture and emerging technologies. *IEEE Access* 2015, 3, 1206–1232.

[3] Sawanobori, T.K. *The Next Generation of Wireless: 5G Leadership in the U.S. (White Paper); Technical Report*; CTIA Everything Wireless: Washington, DC, 2016.

[4] 5G-PPP. *The 5G Infrastructure Public Private Partnership: The Next Generation of Communication Networks and Services. (White Paper).* Available online: https://5g-ppp. eu/wp-content/uploads/2015/02/5G-Vision-Brochure-v1.pdf (accessed on 21 July 2020).

[5] Samsung. *5G Vision (White Paper); Technical Report*; Samsung Electronics Co.: Suwon-si, 2015.

[6] Zanella, A., Bui, N., Castellani, A., Vangelista, L., and Zorzi, M. Internet of things for smart cities. *IEEE Internet Things J.* 2014, 1, 22–32.

[7] Nassar, A.S., Montasser, A.H., and Abdelbaki, N. A survey on smart cities' IoT. In *Proceedings of the International Conference on Advanced Intelligent Systems and Informatics, Cairo, Egypt, 9–11 September.* 2017.

[8] Yan, J., Liu, J., and Tseng, F.M. An evaluation system based on the self-organizing system framework of smart cities: A case study of smart transportation systems in China. *Technol. Forecast. Soc. Chang.* 2020, 153, 119371.

[9] Peng, M., Sun, Y., Li, X., Mao, Z., and Wang, C. Recent advances in cloud radio access networks: System architectures, key techniques, and open issue. *IEEE Commun. Surv. Tutorials*, 2016, 18, 186.

[10] Faisal Ahmed Al-Nasser, M. S. M. *Wireless Sensors Network Application: A Decentralized Approach for Traffic Control and Management*; IntechOpen: Rijeka, 2012.

[11] Pandit, K., Ghosal, D., Zhang, H. M., and Chuah, C.-N. Adaptive traffic signal control with vehicular ad hoc networks, *IEEE Transac. Vehicular Technol.*, 2013, 62, 1459.

[12] Gottlich, S., Herty, M., and Ziegler, U. Modeling and optimizing traffic light settings in road networks, *Comput. Operations Res.*, 2015, 55, 36.

[13] Uchida, N., Tagawa, T., and Sato, K. Development of an augmented reality vehicle for driver performance evaluation. *IEEE Intell. Transp. Syst. Mag.*, 2017, 9(1), 35–41.

7 Cyber-Physical Security Issues and Challenges Using Machine Learning and Deep Learning Technologies

Sima Das, Ajay Kumar Balmiki,
and Kaushik Mazumdar
Maulana Abul Kalam Azad University
of Technology, West Bengal

Parijat Bhowmick
Indian Institute of Technology, Guwahati

CONTENTS

DOI: 10.1201/9781003248750-7

7.1   INTRODUCTION

Today's world is emerging in technology in such a way that back then we only used to think the things are possible or not but now it's really just like magic, sitting in one place and one can do the task or it performed in another place. All credit goes to digitization techniques which are making technology ultimate advance and become paramount in this digital world. Monitoring and controlling of systems became very easy when the Internet of Things (IoT) was introduced to us. IoT generally means connection between hardware (device) and the Internet. IoT is having a wider application in everywhere and every place such as agriculture, healthcare and robotics. IoT is taking our world into the next step, and the main benefit of the IoT system is that it can be assembled as per need. This also proves that it is cost-efficient though one can assemble in his/her way. IoT is just like a big digital tree, and its roots spread all over the world. It will become tougher or next to impossible if IoT is absent for one day from our world. IoT has got such a limelight and fabricated itself in home automation such as lighting equipment, cameras for security, equipment for heating, remote monitoring and controlling and other appliances. The best companions for IoT are cloud, edge and fog technologies. Cloud brings inundation in our digital world; the technology rises to its peak and hence introduces lots of things which are possible nowadays. Back then monitoring and controlling was next to impossible and also the storing of data other than personal devices, presently this all is easy as bean. When talking about another computing technique such as fog, we really have to give some credit little more than cloud because fog has solved many problems which cloud could not. The main difference between these two computing techniques is that fog is decentralized and cloud is centralized, and thus, fog has a tendency of huge communication when it comes to device communication, and hence, one more major problem was bandwidth is also solved in fog as well. Edge also contains security threats regarding data. Several methods like machine learning (ML) and deep learning (DL) are widely used in our industry. These methods play an important role in digitalization. If we look into robotics or any type of advanced system, there must be the concept of ML [1–3].

The rest of the chapter is as follows: Section 7.2 contains literature survey; Section 7.3 is holding overview of IoT techniques used for security; in Section 7.4, cloud, edge, fog techniques has been discussed briefly; ML methods used for security has been covered in Section 7.5; Section 7.6 concludes DL methods used for security; security application areas are included in Section 7.7; and the last one is conclusion.

7.2   LITERATURE SURVEY

In this section, we provide a literature survey of security on cyber-physical system (CPS).

Ni et al., experiment of various data and specialized gadgets sent in savvy networks. The digital protection is a major concern among various organizations across the globe. This chapter initially presents conceivable digital assaults that the perception framework and control framework in shrewd lattice might face, and afterward presents guides to delineate the effect of such assaults. Following it up, a network protection astute guard framework is proposed which will prepare framework administrators an amazing asset to effectively battle programmers and keep the force framework secure and stable. The edge work and the execution procedures of this safeguard framework are additionally introduced [4].

Slipachuk et al., the cycle model of the Basic Framework Network Protection the board utilizing the Incorporated Arrangement of the Public Digital protection Area the executives in Ukraine has created and introduced it. The useful parts of the digital protection arrangement of basic framework articles have been portrayed. The applied worth of the coordinated administration framework for dealing with the network safety of basic foundation objects has been validated [5].

From Wu et al., network safety is one of the main dangers for a wide range of digital actual frameworks (CPS). To assess the digital protection hazard of CPS, a quantitative hierarchies appraisal model comprising assault seriousness, assault achievement likelihood and assault outcome is proposed, which can evaluate the danger brought about by a continuous assault at have level and framework level. Then, at that point, the definitions and estimation techniques for the three records are talked about exhaustively. At last, this chapter gives the danger evaluation calculation which portrays the means of execution. Mathematical model shows that the model can react to the assault opportunity and acquire the framework security hazard change bend. With the goal that it can help the client's reaction to the danger ideal, the danger change bend can likewise be utilized to foresee the danger for the future time [6].

In their paper, Pavlenko and Zegzhda proposed another way to deal with security assessment for cyber-physical frameworks. Existing security techniques appropriate to data frameworks are incapable for cyber-physical frameworks, which comprise data and actual parts. Creators detail prerequisites for security assessment considering key attributes of cyber-physical frameworks. Investigation of logical examination committed to improvement of a security appraisal for cyber-physical frameworks has shown their irregularity with the detailed prerequisites. Creators propose another evaluation of cyber-physical framework's security as digital sustainability, which guarantees maintenance of framework in a steady consistent state under damaging impact. Creators additionally propose a computational measure for surveying digital manageability, in light of self-closeness evaluation. As a mathematical measure, the Hurst type was picked. Let trial tests exhibited adequacy of proposed approach and high affectability of the Hurst example to ruinous impact on cyber-physical frameworks [7].

Barrère et al. observing frameworks are fundamental to comprehend and control the conduct of frameworks and organizations. Digital actual frameworks are especially fragile under some viewpoint since they include continuous imperatives and actual marvels that are not normally thought to be in like manner IT arrangements. Subsequently, there is a requirement for openly accessible checking instruments ready to think about these perspectives. In this banner/demo, we present our drive, called CPS-MT, towards an adaptable, continuous CPS checking instrument, with a specific spotlight on security research. We first present its design and primary parts, trailed by a MiniCPS-based contextual analysis. We additionally portray a presentation investigation and starter results. During the demo, we will talk about CPS-MT's abilities and limits for security applications [8].

Teoh et al. proposed probabilistic models that can be utilized for determining time series information. It has seen accomplishment in different spaces like money, bioinformatics, medical care, farming and counterfeit intelligence. Notwithstanding, the utilization of Well in digital protection found to date is numbered. We accept the properties of Well being prescient, probabilistic, and its capacity to demonstrate distinctive normally happening states that structure a decent premise to display network protection information. It is henceforth the inspiration of this work to give the underlying consequences of our endeavours to anticipate security assaults utilizing Well. A huge organization dataset addressing network safety assaults have been utilized in this work to build up a specialist framework. The attributes of attacker's IP locations can be removed from our incorporated datasets to create measurable information. The network protection master gives the heaviness of each quality and structures a scoring framework by commenting on the log history. We applied Gee to recognize a network protection assault, uncertain and no assault by initially breaking the information into three groups utilizing fluffy k-mean, then, at that point, physically marking a little information (expert instinct), and lastly used Gee state-based methodology. Thus, our outcomes are exceptionally uplifting in contrast with discovering peculiarity in a network protection log, which for the most part brings about making enormous measures of bogus discovery [9].

Chu et al. proposed a security appraisal plot for related digital actual force frameworks. The focal point of the plan is to evaluate the effect of hub disappointments in the correspondence network on the absolute force framework. A possibility set is shaped by disappointments of every hub in the correspondence organization, which is positioned by the hub significance. Two instruments to trigger falling disappointments of the force network are considered, including re-directing of force stream and secret disappointments of insurance following the underlying arbitrary occasion. After the possibility set is listed, the N-1 security evaluation for the reliant digital actual force framework is finished. The effect of every possibility is estimated by the subsequent burden loss of the force framework. Re-enactment after-effects of New Britain 10-machine 39-transport test framework exhibits the adequacy of the proposed conspire [10].

As per Elfar et al., CPSs are generally administrative control frameworks where a human-on-the-loop (HOL) manages at least one independent framework, while installed independence permits the administrators to irregularly take care of the framework and different undertakings. Subsequently, it is basic that the plan of any security-mindful CPS considers the effect of the human association with the framework on security. However, there has been almost no work on the plan of human-CPS that advances human situational mindfulness for upgraded framework execution, especially as far as digital actual security and constant protection against attack. One of the primary obstructions to the quick headway of this field is the shortage of test beds for assessing security-mindful human-CPS communications. We present RESCHU-SA, an extendable virtual stage that works by concentrating on the effect that HOL has on security of CPS with differing levels of independence. It permits clients to examine how inductive thinking and capacity to give settings, especially during an attack, influence the general CPS security. The proposed stage is an augmentation of the Exploration Climate for Administrative Control of Heterogeneous

Automated Vehicles (RESCHU) re-enactment climate, recently utilized in different applications including concentrates on zeroed in on administering automated airborne vehicles (UAVs) missions and assessment of interface ease of use [11].

Craggs and Rashid et al. are intricate foundations containing different innovative antiquities, creators, administrators and clients. Existing exploration has set up the security challenges in such frameworks just as the job of usable security to help people in compelling security choices and activities. In this chapter, we centre around savvy digital actual frameworks, for example, those who are dependent on the Web of Things (IoT). Such brilliant frameworks mean to keenly mechanize an assortment of capacities, determined to conceal that intricacy from the client. Besides, the communications of the client with such frameworks are more regularly verifiable than express, for example, a walker with wearable strolling through a brilliant city climate will in all probability associate with the shrewd climate certainly through an assortment of deduced inclinations dependent on recently gave or consequently gathered information. The key inquiry that we investigate is that of engaging computer programmers to logically consider how clients settle on educated security decisions about their information and data in a particularly inescapable climate. We talk about a scope of existing systems thinking about the effect of mechanization on client practices and contend for the need of a shift from convenience to security ergonomics as a key prerequisite when planning and executing security highlights in brilliant digital actual conditions. Obviously, the contemplations apply more comprehensively than security; however, in this chapter, we centre just around security as a key concern [12].

Shichkina and Fatkieva proposed a strategy for data security; the executives have been created utilizing a primary parametric union of models of digital actual frameworks dependent on piecemeal-linear aggregates. The technique permits you to reconfigure components of complicated specialized articles utilizing the interface administrator [13].

Teoh et al. proposed recurrent neural networks (RNNs), which is an uncommon class of profound learning calculations utilizing neurons or hubs, and have gotten a lot of consideration in the subject of information science in the new year. In RNN, the info hubs think about the current data sources, yet the recently seen yields also – thus the term recursive. In today's unique circumstances, cell phones are just like a piece of cake; pretty much every individual concludes their day-to-day routines. The interest, improvement and utilization of Android gadgets are huge. As Android gadgets rule the current piece of the pie, the topic of safety normally emerges in our complicated world. Thus, the measure of malware information accessible for research is voluminous also. This distribution shows the force and effectiveness of RNN applied onto Android malware information. We study a secured dataset, with more than 4,000 sections named as malignant or harmless. From our investigation and information examination, we present an expectation precision of 0.964 utilizing RNN [14].

7.3 OVERVIEW OF IOT TECHNIQUES USED FOR SECURITY

The IoT expresses the physical devices that are linked to the Internet; these physical devices are situated everywhere all around the world. This is all means to share and transfer the data as per our need. This works for acquiring data from sensors and transfer to the respective sector. At first the sensor collects the physical data from the environment or from the respective places and then transfers into the devices such as microcontroller or microprocessor, but to execute the function of microcontroller or microprocessor, we need IoT. The main part of IoT is to connect the device to the Internet to acquire the real-time data without any effort or presence of human beings. We don't need that much man-power for this process; we can also say that this process can be done without including humans or in the absence of humans. With the help of IoT, we can send data to the cloud, or it also helps in processing data in real-time execution. In the world of cloud computing and digital intelligence, one can deftly develop wireless networks and IoT also gives the prospect to change the things from needle to sword in terms of technology.

7.3.1 ADVANTAGES OF IoT

There are lots of benefits using IoT in our daily life as well as technology and data intelligence, though cloud has also some benefits using IoT. Some of the advantages of IoT are shown in Figure 7.1.

1. **Data Monitoring**
   Efficiently one can monitor the data as well as control the data using IoT systems in digital intelligence. No extra manpower is needed for control-ling and monitoring the data; all we have to do is to give the responsibility to the system and then after the work and action will be done remotely just like an irrigation system. Farmers can control and monitor the plant by their home; they don't have to go to the fields for watering the plant and cash crops.

2. **Easy to Access**
   Accessing real-time data is easier using IoT, talking about the real-time data, location details and online bookings system can be done remotely. Back then it was not so easy to track the real-time data, but nowadays IoT makes it simpler and easier. Life has become so easy with this advanced technology that by sitting in another country, one can access the data of another country.

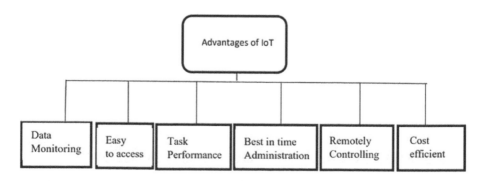

FIGURE 7.1 Advantages of IoT.

Cloud computing plays the main role in it; data can be sent in the cloud, and one can also access data from the cloud easily then and there.

3. Task Performance

IoT also makes sure multiple operations take place faster and better. For example, in agriculture and healthcare purposes, in this type of field, we need the speed to save time and effort. Digitalization in industry can only grow faster due to the performance, and hence, IoT furnishes the speed.

4. Best in Time Administration

Due to the digitalization in the industry and development of IoT, we can save our time very well. For example, if we want to read a newspaper, we don't have to go to the shop and buy it. We can have it on our phone easily.

5. Remotely Controlling

The best examples will be healthcare and agriculture. In agriculture, farmers used to collect the data, and by the help of the data, they used to give the necessary operation to control the parameter or this operation can also be performed remotely. Farmers don't need to go to the field regularly. From home, they can remotely control their plants. And this can become only possible with the help of IoT.

6. Cost-Efficient

Again, for the value of money, IoT manifested itself as a champion. Only with the help of IoT, one can assemble its system as per their needs, so this makes the system more cost-efficient and easy to handle.

7.3.2 Limitation of IoT

Here we discussed some important limitations in IoT which need to be managed and which are very important in digital intelligence. The disadvantages of IoT are shown in Figure 7.2.

1. Infringement of Data

Accessing data easily is quite good enough but accessing data brings some security threat also; it makes our personal data vulnerable and Black Hat people can hack it. Most companies and clients get affected by this problem; this standard is something that cannot be guaranteed, so an untrustworthy issue might occur.

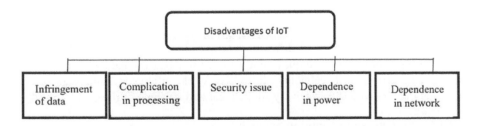

FIGURE 7.2   Limitation of IoT.

2. Complication in Processing

It may look like that IoT can administrate the operation very easily, but there is a lot of complicated peculiarity behind that system. Lots of situations can arise when the system error occurs, and this error may occur by human or nature, so that the situation becomes complicated to handle. If we take the example of an irrigation system, suppose in the field, somehow the water gets inside through the system and then the system starts malfunctioning, so this type of complication is much tougher to manage.

3. Security Issue

Digitalization in technology must be a good thing, but it also brings some security threats, and these security threats can become more baleful for us. Our private data become vulnerable. If we take the example of the healthcare system, suppose a Black Hat person hacked your prescription and changed the medication, obviously you will be affected by that. Just like this, lots of conditions are there. So, most of the time, IoT becomes a security threat.

4. Dependence in Power

Though we all know every system needs some power. There are a huge number of devices which are interconnected to each other, and there are several factors which are dependent. To fulfil all these needs or to access the system, we need an efficient power continuously. If not power then we need some backup devices; if backup devices are not used then it might fail to process the system.

5. Dependence in Network

When it comes to digitalization or digital intelligence, the only word which comes to our mind is Internet. Without the help of the Internet, digital intelligence cannot proceed. So, the Internet plays a very important and vital role in this. There is also one other important fact that is cloud. When we want to access the data from the cloud, we need the Internet and when we want to send the data in the cloud then also we want the Internet. There are various devices which are connected to a certain network or a global network, for transferring the data uninterrupted. This can be wired as well as wireless communication.

7.4 CLOUD, EDGE, AND FOG TECHNIQUES

In this section, we will discuss the role of cloud, edge and fog techniques for CPS and relation between three types of computing, as shown in Figure 7.3.

7.4.1 CLOUD

In this section, we discuss the different types of cloud techniques. To discuss the different types of cloud techniques, first we need to acknowledge the cloud service model. Digitalization of technology in today's world infatuated humans in such a way that they try to scrutinize every single facet and attribute of the cloud.

7.4.1.1 Advantage of Cloud in Security

Advantages of cloud securities are as follows:

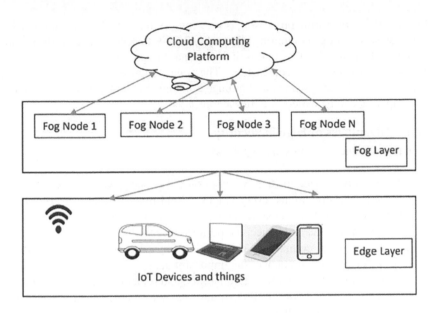

FIGURE 7.3   Understanding of cloud, edge and fog computing.

1. Security of data is one of the main concerns; nowadays data breaching is very common so to protect the data, Cloud security uses high protocol and maintains relationships.
2. The second one is distributed denial-of-service attack which is everybody's concern, but not to worry, in this case, the cloud security also serves us the better toil to reduce the effect or attack which can sabotage our data.
3. Authoritative acquiescence is also better in cloud security; it gives 24×7 support for security purposes and reduces the risk of personal data loss or data vulnerability. This is also beneficial for the company.

7.4.1.2   Limitations of Cloud in Security

Disadvantages of cloud securities are as follows:

1. Issues of bandwidth arise sometimes; both the parties (client and provider) get affected by this type of issue.
2. Transferring of data can cause a serious problem, capacity problems may arise from the client side and this only causes the data error.
3. Sometimes a security threat is also there for the network and also company's employee may use client private data for their own purpose.

7.4.2   EDGE

IoT has set a new precedent in the world of technology; when it comes to IoT, the best companion for IoT is edge computing because it is a distributed computing technique.

In IoT, real-time computing processing is very important, and edge can do it in a better and fastest way. Edge computing allows the devices to access the data storage and computation from nearby. Due to this, it doesn't have to go through latency problems and thus the performance becomes good. The edge computing is fabricated because of IoT only; in IoT, data is generated in huge amounts. So, to tackle all these points, edge computing can be called as a life saviour.

7.4.2.1 Advantages of Edge in Security

Advantages of edge securities are as follows:

1. Edge already sets its position when it comes to speed; when working with real-time data operations, every single second matters; due to traffic, the performance can become slow. This problem is also solved by edge, as it manages the traffic, and hence, the performance will increase.
2. The issue of local compliance is also solved by edge; not only this, it also helps to support the privacy of the important data
3. When it comes to deal with cost, reliability, and scalability, edge has a good performance in it. It's cost-efficient and manages data in such a way that everything becomes reliable.

7.4.2.2 Disadvantages of Edge in Security

Disadvantages of edge securities are as follows:

1. The edge storage problem is fabricated; it requires more data storage when it comes to deal with IoT, and also maintenance is high in edge because the system is distributed, which simply means that there will be a huge cluster of machines present in the network.
2. Data breaches might occur here, and this chance becomes higher when any IoT device comes under the network, though data trespassing can occur in this case.
3. In edge, incomplete data execution is done partially, which means that it only selects half set of data, and this discarded data can be vulnerable and tarnish the security of the user.

7.4.3 Fog

Fogging, fog network and fog computing are the same; the main difference between cloud and fog is that decentralization and flexibility are more in fog than cloud. The fog is better in security in such a way that it gives its supplementary firewall support to protect the user.

Fog networks generally appreciate the IoT logic, for example, if we take agriculture or healthcare, their main functionality is using IoT, and consequently, this may be done by humans like controlling and monitoring etc.

7.4.3.1 Advantages of Fog in Security

Advantages of fog securities are as follows:

1. Data security is better in fog than cloud because you don't need to send the data to a centralized cloud and hence can be managed easily by the IT team. It also has the tendency to connect multiple devices in the similar network; due to this process other than centralized cloud process, it becomes easier to identify the security threat and can be managed before doing any damage.
2. When the factor feasibility comes in our mind then fog is better, because a client can use fog service and create its own product as they want. Though it is very helpful for programmers also because programmers use various types of tools, they can easily do it here.
3. Bandwidth problem is also solved in fog; the data transmission is done by the help of bandwidth, and this costs a lot, so to solve this issue, fog has been introduced; we can send it locally by the help of IoT devices instead of sending the information or data in the cloud.

7.4.3.2 Disadvantages of Fog in Security

Disadvantages of fog securities are as follows:

1. Due to its complex features and logic, it only makes the network more complex and hence becomes more adamant to acknowledge.
2. Though it has a large number of devices in this architecture, there is a huge chance of man-in-the-middle attack, and hence, due to the large spectrum of network and devices, it creates less secure circumambience.
3. Maintenance in fog is not as easy as in the cloud. Just because a huge number of devices are disseminated, it is decentralized and tougher to maintain.

7.5   ML METHODS USED FOR SECURITY

The role of ML in cybersecurity is very important aspect for today's researchers [15]. In Table 7.1, IoT security-related ML methods are discussed.

7.6   DL METHODS USED FOR SECURITY

In this section, IoT techniques with DL for security are discussed in Table 7.2.

7.7   SECURITY APPLICATION AREAS

In this section, we discuss securities and application areas of CPS as shown in Figure 7.4.

a. Healthcare

Healthcare systems nowadays are growing so fast and digitalization in this field even makes us smarter. That's why at present we used to call it the "Smart Healthcare System". This is developed for the management of records for the ease of doctors, patients and other managing staff.

TABLE 7.1
IoT Techniques with ML for Security

Methods	Advantages	Disadvantages	Application in IoT in Security
DT (decision tree)	During pre-processing, DTs require minimum effort compared to another algorithm. DT does not need any normalization and scaling data. It does not affect the missing value. Last but not the least, the process is easy to access for stakeholders	DT method is not stable. The minimum changes of data can affect the whole data structure	Smart healthcare [16]
SVM (support vector machine)	SVM is efficient for higher dimensional spaces. SVM is a memory capable technique, and works well when classes are clearly separated	SVM is not suitable for huge amounts of datasets or big data. There is a problem of classification when target classes are overlapping and another problem is when features dataset is greater than training dataset	Network intrusion and cyber-attack detection [17]
NB (naïve Bayes)	The NB works quickly; the technique is suitable for solving multi-class prediction problems. The method is more effective on categorical input variables than numerical variables	Limitation of this method is not applicable for all real-world problems because it assumes that all features are independent in nature	Brute force attack visualization [18]
kNN (k-nearest neighbours)	In the kNN algorithm, any new data can be added during any stage of the process seamlessly. kNN method is an easy to implement method	Limitation of the kNN method is not working properly for big and high dimensional data. The method is very sensitive in nature for noisy data and missing values	Encrypted uncertain data for Cloud-IoT ecosystem [19]
AR (augmented reality) algorithm	AR can conduct with the real world simultaneously. The method enhances perception and can be used by anyone. The method can help the end user to make a decision	Lack of privacy is the main problem of using AR	Intelligent smart spaces [20]

(Continued)

TABLE 7.1 (*Continued*)
IoT Techniques with ML for Security

Methods	Advantages	Disadvantages	Application in IoT in Security
k-Means clustering	k-Means clustering is simple to implement, good for large amounts of data and easily adoptable	The limitation of this method is that the method does not work well within the global cluster	Anomaly detection in network traffic [21]
ICA (independent component analysis)	It is the signal processing method which separates a multivariate signal to supplement components	The limitation of the ICA is the polarity of the projected component may not be reliable	Enhancing security of ICA-Atom [22]
Multilayer perceptron	Multilayer perceptron's can be applied to a large number of data; it can be suitable for complex nonlinear problems. The technique is very fast to predict output	The predicting value of the method depends on the training dataset	Intrusion detection for CPS [23]
Back propagation neural network	The back propagation neural network is a very easy and fast process to predict output. The method is a very efficient standardized method that does not require any special features for input	The limitation of this method is that the predicted value depends on input value. This technique is not suitable for noisy data	Clone detection [24]
Random forest	Random forest technique can solve classification and regression-related problems. Missing value problems can be automatically handled by random forest technology	Limitation of random forest technique is to take a long time for training a dataset	Cybercrime detection [25]

TABLE 7.2

IoT Techniques with DL for Security

Methods	Advantages	Disadvantages	Application in IoT in Security
CNN (convolutional neural network)	CNNs can store information on the whole network. The CNN technique is able to work with insufficient information. It is distributed in nature	The limitation of CNNs is that it is machine-dependent	Malware detection [26], knowledge tracing [27]
RNNs	RNNs can remember each information through time; this is called long short-term memory (LSTM). LSTM is useful because all inputs and features are stored on the network	Limitation of RNN is that training to the network is a very difficult task	Secure and control DC microgrids [28], threat analysis [29]
GAN (generative adversarial network)	GAN is the technique which can easily detect and recognize any object. GAN can interpret into many versions so it can easily mesh with any ML technique	To train a GAN model is very difficult. The GAN technology is very complex in the field of text or speech recognition	Anomaly detection [30], intrusion detection [31]

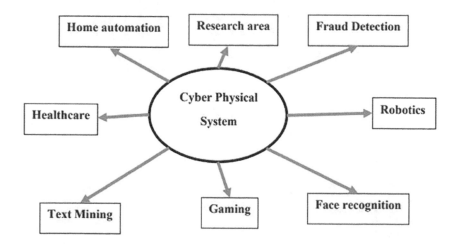

FIGURE 7.4   Application areas of CPS.

b. Home Automation

Home automation systems are what we use generally in homes such as remote sensing, monitoring and controlling; even the best example of security in home automation is cameras, as we use cameras for our own needs and our own security purpose.

c. Fraud Detection

Fraud detection helps to manage illegal activities. Basically, it is a set of protocols and processes that allow the user to keep their identity, data, files, information and many more from the intruders or hackers. These hackers are also termed as Black Hat people. There are many types of attacks which can be done by the hackers such as man-in-middle attack and this attack is very popular, denial-of-service attack, data spoofing, data transit attack, eavesdropping and phishing. These attacks are more than enough to steal important things like personal data from us and leave us vulnerable. So, to tackle the things, fraud detection has been introduced to us and hence provides us with security from intruders. Now this is really possible with the help of ML.

d. Text Mining

Text mining is the process of changing unstructured data into structured data format. Unstructured data means the data which does not contain any predefined set of values or data; it's really hard to process, and structured data is a standardized set of data and consists of various row and column such as name and address. It is used to process the ML algorithm and also it helps to detect the pattern of unstructured data by applying various DL and ML algorithms.

e. Face Recognition

If we come to the authentication technique, one of the most advanced and famous techniques has been introduced to us; the face recognition technique. It is used in authentication, and this authentication is done by the matching of a human face with the provided data set of faces from a digital image. DL has a huge variation and flexibility for this type of complex system technique, and this technique is used in real time [32].

f. Gaming

In today's world, everyone is crazy about gaming, and the gaming industry is earning huge amounts of profit in the market. But the deal is just not over here yet, although security is also concerned, the gaming software is also not secured and can be slayed by the hackers any time. Online gaming can be easy to hack because there is a real-time interaction with the help of a network, and hackers might use this opportunity and can hack your system easily if your system is not secure enough to manage this [33].

g. Robotics

Robotics is a very important field in technology, as without this field or without advanced technology, no one can stand in this world. This technology is growing faster, and it also has a wider application in the digital world. The best companion of robotics is IoT, cloud, ML, artificial intelligence and DL. Different methods have their distinct features. Not only these but also various algorithms are also used in this field to make the system or machine as advanced as possible. IoT has an advanced mechanism which can keep us secure, and through ML and DL, it is quite possible to achieve security due to their distinct algorithms and methods.

7.8 CONCLUSION

The most important factor in cloud computing and in any system is security. Today's digital world is very much concerned about security and its repercussions. This challenge is taking limelight vigorously day by day. Cloud, fog and edge are the new technologies which are mostly used in this digital world. Majorly companies and digital industries are getting affected. Lots of methods, algorithms and techniques have been introduced to us but still there is something we cannot call a fully secure system yet. Now also somewhere this situation arises. In this chapter, we have successfully discussed the techniques, algorithms and methods in cloud for security. Digitalization is good for our generation and future, but when there is advantage, disadvantage come along with it, and this disadvantage is one of major problems to all of us and also it comes with some limitations. The new technologies such as IoT, ML and DL are also used in computing generally when it comes to IoT; in these technologies, security threat also becomes a major problem, and hence, we use some algorithm, authentication techniques for security. In the case of IoT, if the data is hacked then the system will get badly affected, and hence, it can also make personal data vulnerable. This security threat is also the same for ML and DL techniques. But digitization in technology is helping very much and it also has a great future ahead, so there will also be chances in future to discover new methods and technologies which give us full proof security.

REFERENCES

[1] S. Das, L. Ghosh, and S. Saha, "Analyzing gaming effects on cognitive load using artificial intelligent tools," *2020 IEEE International Conference on Electronics, Computing and Communication Technologies (CONECCT)*, 2020, doi: 10.1109/CONECCT50063.2020.9198662.

[2] S. Das, and A. Bhattacharya, "ECG Assess heartbeat rate, classifying using BPNN while watching movie and send movie rating through telegram," In: Tavares, J.M.R.S., Chakrabarti, S., Bhattacharya, A., and Ghatak, S. (eds.), *Emerging Technologies in Data Mining and Information Security. Lecture Notes in Networks and Systems*, 2021, vol. 164, pp. 465–474. Springer, Singapore. doi: 10.1007/978-981-15-9774-9_43.

[3] S. Das, J. Das, S. Modak, and K. Mazumdar. (2022). "Internet of Things with machine learning based smart cardiovascular disease classifier for healthcare in secure platform," In: Bhattacharya, B., Roy, B., Sur, S.N., Mallik, S., and Dasgupta, S. (eds.), *Internet of Things and Data Mining for Modern Engineering and Healthcare Applications*, pages 20. Chapman and Hall/CRC, New York.

[4] M. Ni, A. K. Srivastava, R. Bo, and J. Yan, "Design of a game theory based defense system for power system cyber security," *2017 IEEE 7th Annual International Conference on CYBER Technology in Automation, Control, and Intelligent Systems (CYBER)*, 2017, pp. 1049–1054, doi: 10.1109/CYBER.2017.8446449.

[5] L. Slipachuk, S. Toliupa and V. Nakonechnyi, "The process of the critical infrastructure cyber security management using the integrated system of the national cyber security sector management in Ukraine," *2019 3rd International Conference on Advanced Information and Communications Technologies (AICT)*, 2019, pp. 451–454, doi: 10.1109/AIACT.2019.8847877.

[6] W. Wu, R. Kang and Z. Li, "Risk assessment method for cyber security of cyber-physical systems," *2015 First International Conference on Reliability Systems Engineering (ICRSE)*, 2015, pp. 1–5, doi: 10.1109/ICRSE.2015.7366430.

[7] E. Pavlenko and D. Zegzhda, "Sustainability of cyber-physical systems in the context of targeted destructive influences," *2018 IEEE Industrial Cyber-Physical Systems (ICPS)*, 2018, pp. 830–834, doi: 10.1109/ICPHYS.2018.8390814.

[8] M. Barrère et al., "CPS-MT: A real-time cyber-physical system monitoring tool for security research," *2018 IEEE 24th International Conference on Embedded and Real-Time Computing Systems and Applications (RTCSA)*, 2018, pp. 240–241, doi: 10.1109/RTCSA.2018.00040.

[9] T. T. Teoh, Y. Y. Nguwi, Y. Elovici, N. M. Cheung and W. L. Ng, "Analyst intuition based Hidden Markov Model on high speed, temporal cyber security big data," *2017 13th International Conference on Natural Computation, Fuzzy Systems and Knowledge Discovery (ICNC-FSKD)*, 2017, pp. 2080–2083, doi: 10.1109/FSKD.2017.8393092.

[10] X. Chu, M. Tang, H. Huang and L. Zhang, "A security assessment scheme for interdependent cyber-physical power systems," *2017 8th IEEE International Conference on Software Engineering and Service Science (ICSESS)*, 2017, pp. 816–819, doi: 10.1109/ICSESS.2017.8343036.

[11] M. Elfar et al., "WiP Abstract: Platform for security-aware design of human-on-the-loop cyber-physical systems," *2017 ACM/IEEE 8th International Conference on Cyber-Physical Systems (ICCPS)*, 2017, pp. 93–94.

[12] B. Craggs and A. Rashid, "Smart cyber-physical systems: Beyond usable security to security ergonomics by design," *2017 IEEE/ACM 3rd International Workshop on Software Engineering for Smart Cyber-Physical Systems (SEsCPS)*, 2017, pp. 22–25, doi: 10.1109/SEsCPS.2017.5.

[13] Y. A. Shichkina and R. R. Fatkieva, Intelligent information security management of cyber-physical systems, *2021 II International Conference on Neural Networks and Neurotechnologies (NeuroNT)*, 2021, pp. 25–27, doi: 10.1109/NeuroNT53022.2021.9472853.

[14] T. T. Teoh, G. Chiew, Y. Jaddoo, H. Michael, A. Karunakaran and Y. J. Goh, "Applying RNN and J48 deep learning in android cyber security space for threat analysis," *2018 International Conference on Smart Computing and Electronic Enterprise (ICSCEE)*, 2018, pp. 1–5, doi: 10.1109/ICSCEE.2018.8538405.

[15] S. Das, A. K. Balmiki, and K. Mazumdar. "The role of AI-ML techniques in cyber security," In: Prakash, J.O., Gururaj, H.L., Pooja, M.R., Pavan Kumar, S.P. (eds.), *Methods, Implementation, and Application of Cyber Security Intelligence and Analytics*. IGI Global, United Kingdom, 2022. doi: 10.4018/978-1-6684-3991-3.

[16]. R. Manikandan, R. Patan, A. H. Gandomi, P. Sivanesan, and H. Kalyanaraman, "Hash polynomial two factor decision tree using IoT for smart health care scheduling," *Expert Systems with Applications*, 2020, 141, 112924, ISSN 0957-4174, doi: 10.1016/j.eswa.2019.112924.

[17] K. Ghanem, F. J. Aparicio-Navarro, K. G. Kyriakopoulos, S. Lambotharan and J. A. Chambers, "Support vector machine for network intrusion and cyber-attack detection," *2017 Sensor Signal Processing for Defence Conference (SSPD)*, 2017, pp. 1–5, doi: 10.1109/SSPD.2017.8233268.

[18] D. Stiawan, S. Sandra, E. Alzahrani and R. Budiarto, "Comparative analysis of K-means method and Naïve Bayes method for brute force attack visualization," *2017 2nd International Conference on Anti-Cyber Crimes (ICACC)*, 2017, pp. 177–182, doi: 10.1109/Anti-Cybercrime.2017.7905286.

[19] C. Guo, R. Zhuang, C. Su, C. Z. Liu and K. R. Choo, "Secure and efficient K nearest neighbor query over encrypted uncertain data in cloud-IoT ecosystem," *IEEE Internet of Things Journal*, 2019, 6(6), 9868–9879, doi: 10.1109/JIOT.2019.2932775.

[20] J. Purmaissur, A. Seeam, S. Guness and X. Bellekens, "Augmented reality intelligent lighting smart spaces," *2019 Conference on Next Generation Computing Applications (NextComp)*, 2019, pp. 1–5, doi: 10.1109/NEXTCOMP.2019.8883577.

[21] R. Kumari, Sheetanshu, M. K. Singh, R. Jha and N. K. Singh, "Anomaly detection in network traffic using K-mean clustering," *2016 3rd International Conference on Recent Advances in Information Technology (RAIT)*, 2016, pp. 387–393, doi: 10.1109/RAIT.2016.7507933.

[22] W. K. S. Karunarathne and S. P. Wimalarathne, "Enhancing security of ICA-AtoM: The web based archival description software," *2014 First International Conference on Computational Systems and Communications (ICCSC)*, 2014, pp. 236–241, doi: 10.1109/COMPSC.2014.7032654.

[23] P. V. S. Alpaño, J. R. I. Pedrasa and R. Atienza, "Multilayer perceptron with binary weights and activations for intrusion detection of Cyber-Physical systems," *TENCON 2017-2017 IEEE Region 10 Conference*, 2017, pp. 2825–2829, doi: 10.1109/TENCON.2017.8228342.

[24] F. Pan, H. Wen, X. Gao, H. Pu and Z. Pang, "Clone detection based on BPNN and physical layer reputation for industrial wireless CPS," *IEEE Transactions on Industrial Informatics*, May 2021, 17(5), 3693–3702, doi: 10.1109/TII.2020.3028120.

[25] T. Arora, M. Sharma and S. K. Khatri, "Detection of cyber crime on social media using random forest algorithm," *2019 2nd International Conference on Power Energy, Environment and Intelligent Control (PEEIC)*, 2019, pp. 47–51, doi: 10.1109/PEEIC47157.2019.8976474.

[26] C. Chen, S. Wang, D. Wen, G. Lai and M. Sun, "Applying convolutional neural network for malware detection," *2019 IEEE 10th International Conference on Awareness Science and Technology (iCAST)*, 2019, pp. 1–5, doi: 10.1109/ICAwST.2019.8923568.

[27] W. Wang, T. Liu, L. Chang, T. Gu and X. Zhao, "Convolutional recurrent neural networks for knowledge tracing," *2020 International Conference on Cyber-Enabled Distributed Computing and Knowledge Discovery (CyberC)*, 2020, pp. 287–290, doi: 10.1109/CyberC49757.2020.00054.

[28] M. R. Habibi, T. Dragicevic and F. Blaabjerg, "Secure control of DC microgrids under cyber-attacks based on recurrent neural networks," *2020 IEEE 11th International Symposium on Power Electronics for Distributed Generation Systems (PEDG)*, 2020, pp. 517–521, doi: 10.1109/PEDG48541.2020.9244459.

[29] T. T. Teoh, G. Chiew, Y. Jaddoo, H. Michael, A. Karunakaran and Y. J. Goh, "Applying RNN and J48 deep learning in android cyber security space for threat analysis," *2018 International Conference on Smart Computing and Electronic Enterprise (ICSCEE)*, 2018, pp. 1–5, doi: 10.1109/ICSCEE.2018.8538405.

[30] S. K. Alabugin and A. N. Sokolov, "Applying of generative adversarial networks for anomaly detection in industrial control systems," *2020 Global Smart Industry Conference (GloSIC)*, 2020, pp. 199–203, doi: 10.1109/GloSIC50886.2020.9267878.

[31] P. F. de Araujo-Filho, G. Kaddoum, D. R. Campelo, A. Gondim Santos, D. Macêdo and C. Zanchettin, "Intrusion detection for cyber–physical systems using generative adversarial networks in fog environment," *IEEE Internet of Things Journal*, 2021, 8(8), 6247–6256, doi: 10.1109/JIOT.2020.3024800.

[32] S. Sahana, S. Das, and B. K. Sarkar, "Principal component-based method for whole genome phylogenetic analysis without alignment: Application to HEV genotype," in *Computing, Communication & Automation (ICCCA), 2015 International Conference on*, 2015, pp. 984–989.

[33] K. K. Senapati, S. Sahana, and G. Sahoo, "An efficient pattern matching algorithm for biological sequences," in *IPCV*, 2010, pp. 755–759.

8 Brain MRI Image Active Contour Segmentation for Healthcare Systems

Kiranmai Babburu, Gurucharan Kapila,
S S Kiran, and Mallavarapu Rajan Babu
Lendi Institute of Engineering and Technology

CONTENTS

8.1 INTRODUCTION

Magnetic resonance imaging (MRI), a non-invasive and safe technique, creates comprehensive pictures of the brain by using a magnetic field and radio waves to make images of the inside of the skull. If you get an MRI of your brain, it may reveal cysts, tumors, hemorrhages, and edemas, as well as structural issues, infections, and issues with blood vessel health. An MRI of the brain may diagnose chronic headaches, dizziness, weakness, seizures, and chronic nervous system diseases like multiple sclerosis. In some instances, an MRI may give clear images of brain areas that an X-ray, CT scan, or ultrasound cannot.

DOI: 10.1201/9781003248750-8

Teenage brain tumors may range from those healed with little therapy to incurable, even intensive treatment. There are many distinct kinds of juvenile brain tumors like craniopharyngiomas, brainstem gliomas, ependymomas, germ cell tumors, and astrocytoma.

The brain must be segmented into various components such as gray matter, subcortical white, CSF, and the skull to identify tumors, edema, and hemorrhages. According to studies conducted primarily on industrialized nations, people suffering from brain tumors die higher than in the previous three decades. Tumors are clumps of tissue that grow uncontrolled due to mechanisms that usually regulate growth [1]. Tumors can wipe out all of the brain's healthy cells. Other brain regions may be impacted by the crowding and inflammation, edema, and pressure inside the skull that results [2]. In the early medical tumor identification investigations, researchers employed traditional image processing approaches (like edge detection and region growth) depending on the gray intensities of pictures [3].

8.1.1 NEED FOR SEGMENTATION OF BRAIN MRI IN HEALTH CARE SYSTEMS

Computational applications are becoming more critical in everyday life. Computer-aided diagnosis (CAD) tool use in computational biomedical analysis is being studied more deeply, especially in CAD. Brain tumor identification and research are among the most frequently occurring fatalities in today's healthcare environment. As per the National Cancer Institute's 2019 data, the number of cases of carcinogens, including brain cancer, has grown by around 10% over the previous two decades. A total of 29,000 Americans are diagnosed each year with primary brain tumors, and 13,000 of them die as a result, according to the National Brain Tumor Foundation. Particularly in children, one-quarter of all cancer-related fatalities is caused by brain tumors. In the United States, primary brain tumors affect 11–12 persons per 100,000, whereas primary malignant brain tumors affect 6–7 people per 100,000. Over 4,200 British are afflicted with a brain tumor (2007 estimates). Over 200 distinct tumors are discovered every year; mainly in the United Kingdom, 16 out of 1,000 malignancies are found in the brain. However, in India, 80,271 people have various types of tumors.

Fully computerized brain problem diagnosis, such as brain tumor identification using MRI, is a notable example of specialized image analysis in the medical field. Brain tumor develops due to unchecked growth of the brain's tissue cells. Because the cells that provide blood to the arteries are firmly linked together, standard laboratory tests are insufficient to examine the chemistry of the human brain. Three biomedical imaging methods like CT, MRI, and PET allow doctors and researchers to look at the brain's structure without performing invasive surgery.

MRI is a biological imaging technique used by radiologists to study the brain's anatomy. Human soft tissue structural anatomy may be checked in great depth using MRI. The use of an MRI scan may aid in diagnosing a brain tumor [3]. It's utilized to learn human anatomy without going under the knife [4,5]. Brain picture segmentation is one of the essential parts of computer-assisted clinical diagnosis and treatment of tumors and other anatomical irregularities. Noisy brain MR images are multiplicative; thus, reducing noise is vital to getting good results. It is critical to ensure that noise reduction algorithms do not eliminate necessary anatomical features from a therapeutic standpoint. Brain MR image segmentation is, therefore, a challenging research topic. Thus, correct MRI picture segmentation is critical for effective diagnosis utilizing computer-aided clinical techniques. Various approaches for segmenting MR images have been proposed thus far.

Texture analysis or recognition, a subject of the growing interest of computer vision for almost three decades, is one of the most pressing problems. Texture analysis algorithms fall into three categories: statistical, model-based, and structural. Each approach has pros and cons. A tumor is a brain development made up of distinct or mixed cells. Brain tumors are categorized as dangerous or noncancerous (favorable). When noncancerous or harmful tumors grow, they put pressure on the interior of your bone, which may be deadly. Primary and secondary brain tumors are two classifications for brain tumors. A primary tumor begins in the brain. We need a few benign brain tumors. An auxiliary tumor arises when malignant growth cells from another organ, such as the respiratory organ or breast, move to the brain. Nuclear reverberation, often known as X-ray, is a technique used chiefly (but not exclusively) in therapeutic applications to see the inner workings safely. When the body or a portion of it has been placed through a ground-breaking and uniform transformation, it depends on the soothing qualities of electrified synthetic component cores in water. The obvious benefit of this approach is its non-ionizing nature, as opposed to elective procedures like CT outputs and X-ray beams, which expose the patient to ionizing radiations. Image segmentation is often needed to allow experts and specialists to break down the patients' information, such as before medical treatment, to determine the particular region of an organ or a tumor while using MRI.

Clinical identification relies heavily on medical picture segmentation. An ideal clinical picture segmentation characteristics are most minor client association, quick computation, and accurate and consistent segmentation findings. Image segmentation is a kind of picture analysis that attempts to divide an image into several areas based on a homogeneity criterion. The precise information on these divided areas predicts a crucial role in evaluating dangerous malignancies. An enhanced segmentation system for correctly segmenting tumors from Brain MRI image datasets is described in this article, which is submitted to full-scale assortment methods. The approach is based on active contour, which adapts to the form of the item by limiting (among other things) inclination subordinate interest vitality while maintaining the flawlessness of the structure shape. As a result, unlike edge detection, active contour techniques become more noise-resistant as the requirements for form flawlessness and shape soundness work as a kind of regularization.

Another advantage of this method is that prior information on the shape of the article may be combined with the structure definition process. For active contour-based algorithms [6] to converge effectively to the actual boundary, the contour must be initialized near the object boundary.

Because of advances in medical imaging, the number of research automating brain tumor segmentation has exploded in recent decades [7]. The active contour models (ACMs) are used for brain tumor segmentation. These methods are effective for locating the tumor-to-tissue boundary [8]. This approach utilizes criteria borrowed from anatomical and biological information concerning anatomical regions' positions, sizes, and shapes [9]. Curves or surfaces with active contours move in response to weighted internal and external forces. Curve smoothness is determined by internal forces, while curve pushing and tugging toward anatomical targets is determined by external forces.

Early contour determination and leaking in inaccurate edges plague ACMs in general. The bulk of suggested methods for detecting and segmenting brain abnormalities are restricted by extensive computations, lack of complete automation due to brain tumors, and incorrect contour initialization and edges.

8.1.2 IMPORTANCE OF DE-NOISING IN BRAIN MR IMAGES

The process of diagnosis and treatment planning requires the de-noising of medical images. Recently, the pictures have been captured using digital technology. Removing noises from a digital photograph is a difficult task [10]. The noises detract us from the image's natural beauty. The noise should be identified for the de-noising process, and its statistical characteristics should be evaluated [11]. Gaussian, Rayleigh, and impulsive noises are provided during the image capture process. This noise is produced by the sampling and transmission process. It appears as black and white specks in the picture [12,13]. Image de-noising methods include adding a variation filter, altering the domain filter, and using a gradient approach. The impact of smoothing on variation and change domain filters [14,15]. The fuzzy hybrid filter removes Rician noise from MR pictures. Rician noise affects post-processing techniques like segmentation and parametric synthesis [16]. Fair non-local means and a Bayesian estimator may reduce speckle noise in ultrasound pictures. It captures the Gamma distribution's shape and size, as well as each pixel's intensity. A monogenic wavelet depicts noise as a Laplace mixed distribution and a Rayleigh distribution. The non-local mean filter reduces noise by weighing all pixels [17–19].

Wavelet- and curvelet-based transformation filters and statistical filters are anticipated to be presented in future MRI filtering with the particular goal of removing the opacity issue. The bilateral filter (BF) is a well-known nonlinear noise filter designed for MRI in the previous year. It is a simple, local, non-iterative technique for spatial noise reduction and edge preservation. Local information is used in a picture for identifying edge parts. These segments smooth out various portions of the picture at that moment [20]. Despite its widespread usage in MRI image de-noising, nothing is known about the optimal BF settings [21,22]. Statistical methods were used in previous studies to find the best parameters. The authors [23] used vector root mean squared error estimates to adjust the best filter settings. Despite this, their approach acknowledges the existence of a factual image of the world. Another particle swarm optimization PSO-focused optimization research used color images degraded by a single noise level to test the BF proposal's parameters [24]. Most of the time, the BF settings are worked out via a process of try and error. EGOA (enhanced grasshopper optimization algorithm) benefits from the considerable study and quick union time. In this calculation, the remarkable versatility of the tool readily balances inquiry and misuse. This makes the EGOA computation seek an alternative to multi-objective problems.

The multidimensional computational nature outperforms a variety of writing advancement techniques. As a result of these ground-breaking discoveries, we created a multi-objective streamlining agent modeled by grasshopper social behavior. The study aims to enhance the BF parameters by using EGOA to decrease noise in a medical MR image. EGOA has the advantage of finding the best solution faster and needing less prior knowledge about the situation [25]. According to our research, this is the first time a team has used EGOA to optimize the BF parameters used to de-noise MR images.

8.2 LITERATURE REVIEW

In the literature, several segmentation methods for improving segmentation have been described. A list of some of the most often used methods is provided below.

8.2.1 THRESHOLDING

Image segmentation often employs thresholding as one of the techniques. This technique works well with pictures that have varying brightness levels. This technique divides the picture into distinct areas depending on the intensity levels. The segmented image is $Y(x, y)$ if $X(x, y)$ is the input image, and "T" is the threshold value.

$$Y(x,y) = 1, \text{ if } X(x,y) > T$$

$$0, \text{ if } X(x,y) \leq T$$

Using $Y(x, y)$, the picture may be divided into two categories. If we wish to divide a photo into several groups, we'll need multiple threshold points. When we have two threshold values, then $Y(x, y)$ will segment the image into three groups as

$$Y(x,y) = a, \text{ if } X(x,y) > T_2$$

$$b, \text{ if } T_1 < X(x,y) \leq T_2$$

$$c, \text{ if } I(x,y) \leq T_1$$

Bhattacharyya and Taihoon Kim [26] developed an image segmentation method for detecting tumors in brain MRI. In each picture, many current thresholding methods yielded varied results. To get an acceptable impact on brain tumor pictures, they developed a technique for making tumor identification individually. The literature [27–30] provides a review of various thresholding methods.

8.2.1.1 Otsu's Thresholding

The Otsu method is the most widely utilized among the several thresholding methods. Using discriminant analysis, Otsu's thresholding method divides the picture into two classes, $C1$ and $C2$, at each of a set of gray levels "k," with $C1$ being equal to 0 and $C2$, to 1, 2, 3, 5, and so on. This image contains n_i pixels in the ith gray level;

therefore, let P_i be the total number of pixels. The likelihood of a gray level is represented as

$$P_i = \frac{n_i}{n}$$

Classes "$C1$" and "$C2$" indicate the area of interest and the backdrop, respectively.
If the histogram is unimodal or nearly unimodal, the Otsu technique fails.

8.2.1.2 Local Thresholding

For each of the sub-images, a threshold value is determined to determine the quality of the picture. More computation is involved in computing a local limit than a global limit. It achieves good outcomes when used to change the background of photographs. It can only get rid of specks of hair [31].

8.2.1.3 Threshold for Histograms

Histogram thresholding segmentation uses histogram characteristics and gray level thresholding.

The following is the algorithm to be used:

Step 1: A histogram is created for each half of the MRI brain picture split along the image's central axis.

Step 2: The histogram's threshold point is determined by comparing two histograms.

Step 3: The threshold point divides the world into two halves.

Step 4: To determine the physical dimensions of the tumor, the detected picture is clipped along its contour.

Step 5: Make a copy of the original image and compare it to the segmented image. Assuming the value is greater than the threshold, it is given 255.

Step 6: Split the tumor in half.

Step 7: Calculate the size of the tumor.

8.2.2 Segmentation Based on the Edges

Using edge-based segmentation methods, you can split a picture into parts depending on how quickly the intensity changes along the edges [32,33]. A binary image is the end outcome. Gray histogram and gradient-based edge-based segmentation are two of the most often-used methods [34]. In the case of the edge detection technique, the selection of cut-off value T has a considerable influence on the result [35]. The Sobel, canny, Laplace, and Laplace of Gaussian (LOG) operators are used in gradient-based edge detection. The intelligent operator [36] is the most promising.

The detection accuracy and noise immunity of edge detection techniques must be balanced. For example, if noise detection accuracy is set too high, spurious edges may be introduced into the image. This makes the picture outline inaccurate. Certain portions of the image outline may be unnoticed, causing object positions to be incorrect. The use of edge detection techniques is thus appropriate for noise-free and detailed images, but, when dealing with complex and noisy images, they often produce missing or extra edges [37].

8.2.3 SEGMENTATION BY REGION

Algorithms for segmenting the image into regions are simpler and more noise-resistant than edge detection techniques [38,39]. Techniques that use edges to split images into comparable areas based on predefined criteria are called edge-based techniques. In contrast, regions that divide ideas into similar regions are region-based methods.

8.2.3.1 Region Growing

The region growth method is a systematic segmentation strategy. Using this technique, a single seed pixel is used as a starting point, then pixels from the immediate neighborhood are added as required to expand the size of the picture. When a region's growth slows, a new seed pixel that does not belong to any other zone is chosen. When all of the pixels in a given area belong to the same region, the expansion of that region is stopped.

A technique known as area-expanding segmentation is the most effective way to distinguish between small, basic features like tumors and lesions [40,41]. The following are some of the drawbacks of adopting this method: (i) manual involvement is sometimes needed to choose the seed location; (ii) noise-sensitive, resulting in holes or over-segmentation in the extracted areas.

Using homotopic region growth techniques, the discontinuity in the extracted picture may be eliminated.

In paper [6], the authors presented a technique, "Snakes: Active contour models," dependent on crippled shapes and demonstration that isolated the forepart from the backdrop. In the image processing field, these ACMs have been most generally utilized as fruitful systems for image segmentation [6].

Level set strategies are utilized in the models that control the developing contour toward the ideal limits, which are used especially vigorously for complex clinical images analysis.

In the paper, [42] "Active contours without edges," authors have introduced the concept of ACMs, which can be effectively used in image segmenting applications.

In the paper "A deformable segmentation algorithm for brain M.R. images" [43], the authors introduced an algorithm (D-C) for solving segmentation problems effectively. They demonstrated that their algorithm works better by incorporating the active contour segmentation technique changes. They have included shape and measuring images and their energy functions for effective segmentation of brain MRI images.

In this way, Chan-Vese [44] improved the technique and viably portioned the noisy images without smooth limits by consolidating the edge work.

8.3 PROPOSED METHODOLOGY

Segmentation refers to fragmenting the image into different fragments called segments. These segments help in the extraction of regions of interest from the background. This can be done through various methods described in the literature survey.

The covering of tissues, noise, and low complexity lead to inappropriate evaluation of region of interest in the MRI images. The standard edge-based segmentation fails to properly quantify the area of interest due to noise.

8.3.1 Chan-Vese ACM

Chan-Vese model for dynamic forms [45] is an incredible and adaptable strategy that can portion numerous kinds of images, including some that would be very hard to fragment in methods for "traditional" segmentation—i.e., utilizing thresholding or slope-based strategies. The Mumford-Shah [45] practical is used in this model to segment the brain, heart, and trachea. This model [45] is often used in clinical imaging, as shown in Figure 8.1.

Consider Ω to be a bounded open collection containing all possible R^2, with $\partial\wedge$ its boundary. Consider $\mu_0 : \underline{\Omega} \to R$ to be a predetermined image, and $C(s)$ is a piecewise $C^1[0,1]$ defined curve.

The area included inside C will be denoted as $]$, and the space beyond C will be marked as $\dfrac{\Omega}{\omega}$. Furthermore, the value of c_1 will represent the average intensity of the pixels inside C, and the moderate-intensity outside of C will be denoted by the variable c_2 (i.e., $c_1 = c_1(C)$, $c_2 = c_2(C)$).

The goal of the Chan-Vese method is to reduce the energy function [45] to the smallest possible value $F(c_1,c_2,C)$, given by:

$$F(c_1,c_2,C)=\mu \cdot \text{Length}(C)+v \cdot \text{Area}\left(\text{inside}(C)\right)$$
$$+\lambda_1 \int_{\text{inside}(C)} \left|\mu_0(x,y)-c_1\right|^2 dx\ dy \qquad (8.1)$$
$$+\lambda_2 \int_{\text{outside}(C)} \left|\mu_0(x,y)-c_2\right|^2 dx\ dy$$

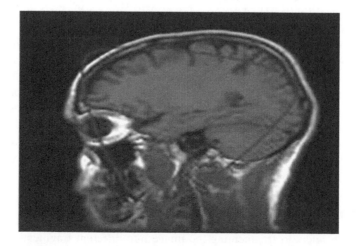

FIGURE 8.1 The image is defined [45] on $\underline{\Omega}$ (the big rectangle), (the arbitrary) red curve is C.

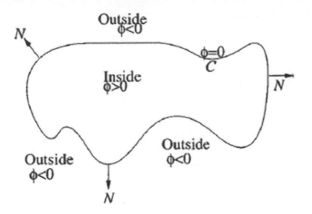

FIGURE 8.2 The function [45] of signed distance $\Phi(x,y)$.

The above equation represents the energy function where $\mu \geq 0, v \geq 0, \lambda_1, \lambda_2 \geq 0$ are settings that cannot be changed.

We rewrite $F(c_1, c_2, C)$ in terms of signed distance function from C as follows:

1. Length (C) is the length of the zero level set function $\Phi(x,y) = 0$ as shown in Figure 8.2

2. Length $(C) - \int_\Omega |\nabla H(\Phi(x,y))| dx\, dy = \int_\Omega \delta\Phi(x,y) |\nabla(\Phi(x,y))| dx\, dy$

   where $H(z) = \{1,$ if $Z \geq 0$
   $\quad\quad\quad 0,$ if $Z < 0$

3. Area $(inside(C))$ the area of the region in which $\Phi(x,y) \geq 0$:

   $$\text{Area}(inside(C)) = \int_\Omega H(\Phi(x,y)) dx\, dy$$

4. $\int_{inside(\cdot)} |\mu_0(x,y) - c_1|^2 dxdy$ can be calculated in terms of $\Phi(x,y)$, when Considering only the region in which $\Phi(x,y) \geq 0$:

   $$\int_{inside(C)} |\mu_0(x,y) - c_1|^2 dx\, dy = \int_{(x,y):\Phi(x,y)\geq 0} |\mu_0(x,y) - c_1|^2 dx\, dy$$

   $$= \int_\Omega |\mu_0(x,y) - c_1|^2 H(\Phi(x,y)) dx\, dy$$

5. Similarly

   $$\int_{outside(C)} |\mu_0(x,y) - c_2|^2 dx\, dy = \int_{(x,y):\Phi(x,y)\geq 0} |\mu_0(x,y) - c_2|^2 dx\, dy$$

   $$= \int_\Omega |\mu_0(x,y) - c_2|^2 H(1 - \Phi(x,y)) dx\, dy$$

6. The average intensities

$$c_1 = \frac{\int_\Omega \mu_0(x,y) H(\Phi(x,y)) dx\ dy}{\int_\Omega H(\Phi(x,y)) dx\ dy}$$

$$c_2 = \frac{\int_\Omega \mu_0(x,y) H(1-\Phi(x,y)) dx\ dy}{\int_\Omega H(1-\Phi(x,y)) dx\ dy}$$

Thus, the energy functional in terms of c_1, c_2, Φ is obtained from the above equations as

$$F(c_1, c_2, \Phi) = \mu \int_\Omega \delta_0 \Phi(x,y) |\nabla\Phi(x,y)| dx\ dy$$
$$+ v \int_\Omega H(\Phi(x,y)) dx\ dy + \lambda_1 \int_\Omega |\mu_0(x,y) - c_1|^2$$
$$H(\Phi(x,y)) dx\ dy$$
$$+ \lambda_2 \int_\Omega |\mu_0(x,y) - c_2|^2 H(1-\Phi(x,y)) dx\ dy$$

Based on the words in the equation above, we may conclude that the curve's development is as follows:

(Two variables had an impact on the outcome. This will be ignored since v is typically set to zero.)

The spherical shape during the evolution process regularizes the curve and smooths it out.

- The "region term" $-\lambda_1 (\mu_0 - c_1)^2 + -\lambda_1 (\mu_0 - c_1)^2$ has an impact on the velocity of the curve
- $\mu \int_\Omega \delta_0 \Phi(x,y) |\nabla\Phi(x,)| dxdy$ is the penalty on the total length of the curve C
- For instance, if the image's borders are smooth, we increase μ to avoid a complicated curve C
- λ_1, λ_2 affect the required homogeneity inside C and outside C, respectively.
- For instance, it would be advisable to set $\lambda_1 < \lambda_2$ wherein the background is dark, and the foreground items are grayscale.

8.3.2 Proposed Model for Segmentation

The proposed methodology, shown in Figure 8.3, involves preprocessing, which helps removing noise through a BF. The segmentation technique used is the ACM.

FIGURE 8.3 Block representation proposed methodology along with flow chart.

Input Image: The input image is taken from various Brain Atlas research webpages containing different datasets of neoplasm in brain MRI.

Preprocessing: The input picture is then de-noised using a BF corrupted with white Gaussian noise of various variances.

Bilateral Transform Noise Filter: The hybrid algorithm blurs a picture while maintaining solid edges. Its ability to decay an image into various scales without causing haloes after the alteration has made it ubiquitous in computational photography applications, such as tone planning, style move, relighting, and de-noising. Its detailing is basic. Every pixel is supplanted by a weighted normal of its neighbors. This is critical since it makes it simple to acquire instinct about its conduct, adjust it to application-explicit necessities, and implement it. BF depends just upon two boundaries that demonstrate the size and differentiation of the highlights to preserve. The suggested solution uses a BF to remove noise while keeping the edges. The de-noised picture is transmitted for active contour segmentation.

Active Contour Segmentation: Segmentation begins with contour initialization, where we select the underlying shape on the image and set the most extreme number of iterations. This algorithm comprises two energies, of which one power diminishes the form and another expansion of the shape. These two energies get adjusted when the form arrives at the furthest point of our ideal item, which brings about total segmentation. The idea of active ACM was first presented in 1987 [6] and was later created by various analysts. A functioning shape is a vitality limiting spline that identifies determined highlights inside an image. It is an adaptable bend (or surface) that can be progressively adjusted to required edges or items in the photo (it very well may be utilized to programed objects division).

The quantity of control characterizes the active contour focuses just as a succession of one another. Fitting active contours to shapes in images is an intelligent procedure. The client must recommend an initial contour, which is very near the planned figure. The contour will, at that point, be pulled into highlights in the image extricated by inside vitality, making an attractor image.

Tumor Extraction: We extract the tumor from the segmented image, our area of interest. Additionally, we remove the tumor's area and the number of pixels inside the tumor's region, which assists doctors in diagnosing and tracking down the neoplasm using brain MRI.

The active contour technique is a numerical and conceptual device created by Osher and Sethian and used in picture processing, liquid mechanics, design, and computer vision, among other domains. Compared to the active contour technique, the active contour strategy offers a few advantages in image segmentation. Topological modifications are no match for the dynamic form approach.

The suggested work contains the following advancements in this calculation.

a. Create an arbitrary network with the exact measurement as the input image, then connect this framework to the picture—this aids in creating the contour in the image. The outline was visible in the image at the time, but this aids in creating a more effective manner.
b. Next, locate the contour location in the picture and create contours to aid image segmentation. This makes the image's first segmentation.
c. Once these contours have been identified in the picture, the following step is to update the various segments by calculating their distance. If the distance is negative, the pixel's value or location is considered part of the segment. If the distance is positive, the pixel's value or place is considered outside the component.
d. The next step is to update the segmented region by examining the segment's pixel values. The function is handled so seriously that any new form may readily accept a change in the area. As the update values change, the ACM moves to merge or divide the segments.

8.4   RESULTS AND DISCUSSION

Preprocessing is utilized to prepare the input brain MRI image, subsequently boosted with additive white Gaussian noise. By sending the signal via the BF, this noise is eliminated. The next stage includes segmentation utilizing the active contour approach, implemented via the Chan-Vese technique.

In this technique, the first mask is selected with the number of iterations. The tumor is split out of the picture. The tumor's area and pixel count are determined. The suggested approach has been tested for validity on a small number of benign tumor MRI datasets across several iterations. In addition to validation, we calculated the general execution of our intended method based on segmentation time, which is one of the most critical factors in the development of fast handing out algorithms, and we presented our results. Our suggested method separated the cerebrum tumor from MRI images in 5 seconds, 7 minutes, 26 seconds, and 7 minutes and 19 seconds, respectively, when everything was normal. The segmentation time of tumors in complete prepared datasets (for example, at 250 iterations) was as low as 5 seconds and as high as 7.9 seconds, depending on the unpredictable clinical picture.

Figure 8.4 shows one of the input images from the brain MRI dataset; Figure 8.5 depicts the image after adding additive white Gaussian noise. Figure 8.6 shows the output of the BF, after which it is segmented using the proposed active contour Chan-Vese technique. Figure 8.7 shows the segmented image where we can observe only the tumor area being segmented out of the whole picture. Finally, Figure 8.8 depicts the neoplasm present in the brain MRI image, which is our region of interest.

Figures 8.4–8.8 are obtained after each step described in the methodology. The above results are generated at a max iteration of 100. The time taken to segment the image is 7.2 seconds. The area occupied by the tumor is 0.0930, with the number of pixels in the tumor area being 6,093.

Figure 8.9 shows one of the input images from the brain MRI dataset; Figure 8.10 shows the image after adding additive white Gaussian noise. Figure 8.11 shows the output of the BF, after which it is segmented using the proposed active contour Chan-Vese technique.

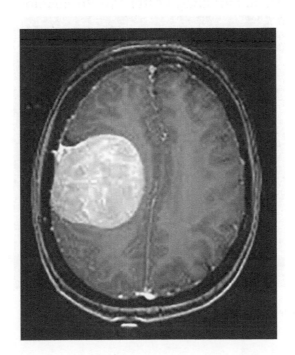

FIGURE 8.4   Input image.

Noisy Image

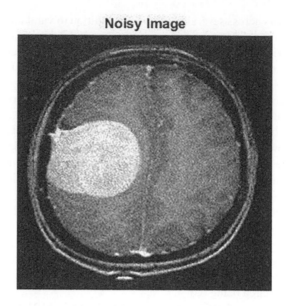

FIGURE 8.5   Noisy image.

Resultant Image After Pre-Processing

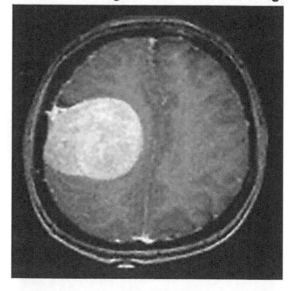

FIGURE 8.6   After preprocessing.

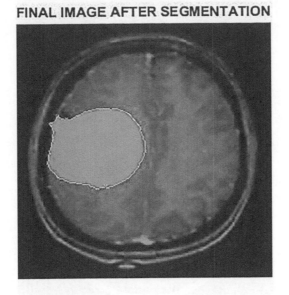

FIGURE 8.7   Segmented image.

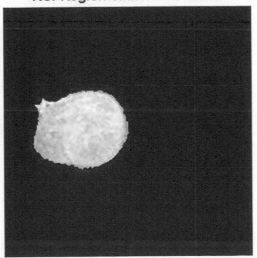

FIGURE 8.8   Neoplasm extracted from brain MRI image.

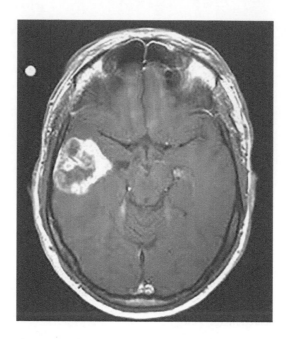

FIGURE 8.9 Input image.

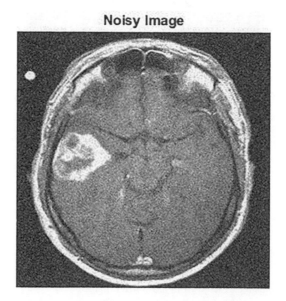

FIGURE 8.10 Noisy image.

Resultant Image After Pre-Processing

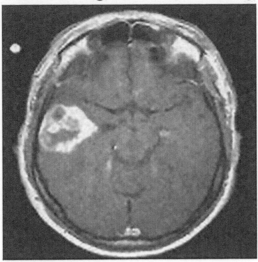

FIGURE 8.11   After preprocessing.

FINAL IMAGE AFTER SEGMENTATION

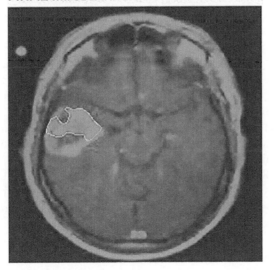

FIGURE 8.12   Segmented image.

Figure 8.12 shows the segmented image where we can observe only the tumor area being segmented out of the whole picture. Finally, Figure 8.13 shows the neoplasm present in the brain MRI image, which is our region of interest.

Figures 8.9–8.13 are obtained after each step described in the methodology. The above results are generated at a max iteration of 35. The time taken to segment the

ROI-Region of Interest--tumor

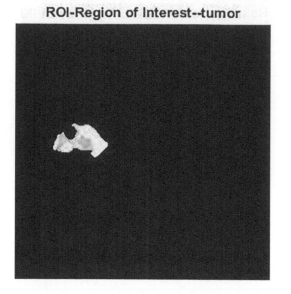

FIGURE 8.13   Neoplasm extracted from brain MRI image.

image is 5.26 seconds. The area occupied by the tumor is 0.0156, with the number of pixels in the tumor area being 1,021. We observe that when the algorithm is run with 35 interactions, we found that the contour does not match the entire tumor area. This problem is called under segmentation which is evident from Figures 8.9 to 8.10.

Figure 8.14 is an example of an input image from the brain MRI dataset; Figure 8.15 shows the image after adding additive white Gaussian noise. Figure 8.16 shows the output of the BF, after which it is segmented using the proposed active contour Chan-Vese technique.

Figure 8.17 shows the segmented image where we can observe only the tumor area being segmented out of the whole picture. Finally, Figure 8.18 shows the neoplasm present in the Brain MRI image, which is our region of interest.

Figures 8.14–8.18 are obtained after each step described in the methodology. The above results are generated at a max iteration of 200. The time taken to segment the image is 4.65 seconds. The area occupied by the tumor is 0.0107, with the number of pixels in the tumor area being 698. We observe that when the algorithm is run with 200 interactions, we found that the contour does not match the entire tumor area. This problem is called under segmentation, clearly evident in Figures 8.17 and 8.18.

Table 8.1 shows the presentation of the proposed framework for various MRI images. It can be observed that the initial contour changes from image to image depending upon its complexity. The max iterations required for segmenting the image also vary from image to image. It can be noted that images with less tumor size (S. no. 4) require fewer iterations, and hence, the time taken to segment the picture is less. It can be observed that (S. no. 3) if the number of iterations is more than the optimum value, it results in over-segmentation.

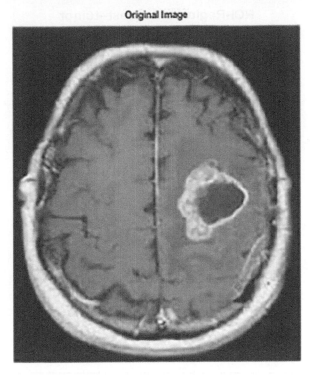

FIGURE 8.14 Original image.

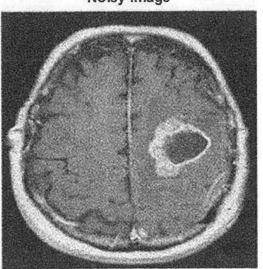

FIGURE 8.15 Noisy image.

ROI-Region of Interest--tumor

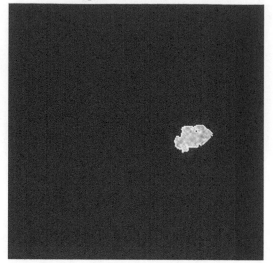

FIGURE 8.16   Preprocessed image.

Resultant Image After Pre-Processing

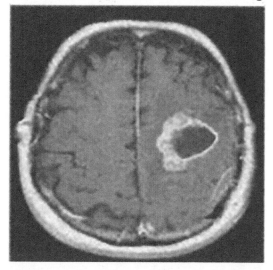

FIGURE 8.17   Segmented image.

FINAL IMAGE AFTER SEGMENTATION

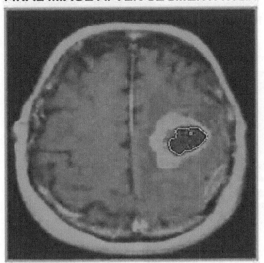

FIGURE 8.18   Neoplasm extracted from brain MRI image.

Table 8.2 shows the performance of the proposed algorithm at different iterations but the same mask. It can be observed that at the 50th iteration, we follow from the segmented image that the neoplasm has not been appropriately segmented. Here observe that as the number of iterations increased from 50 to 250, we see a considerable change in the segmented output. At the 250th iteration, we found that the proposed algorithm could the neoplasm extracted from the MRI image is far better when compared with other iterations.

Hence, we can conclude that perfect segmentation of neoplasm from brain MRI depends majorly on the mask size taken, the number of iterations, and the complexity of the input image. We can also conclude from the above results that our proposed design shows better results in the presence of noise. The BF, which helps in de-noising the image, is the perfect combination for the segmentation of brain MRI images in the presence of noise and finally, and it can be concluded that the results of the proposed design take on an average of about 7.23 seconds for segmentation, which also varies depending upon the image complexity and number of iterations taken to compute the segmentation.

8.5 CONCLUSION

The function of the proposed active contour segmentation is to make available an early beginning judgment on identification, diagnosis of tumor monitoring, and treatment arrangement for the medical doctor—ACM segment images are based on curve flow curvature and contour. ACM is utilized for prospect analysis and processing. This chapter provides an exceptional and robust technique for segmenting brain tumors from MRI images using the Chan-Vese segmentation method. In this methodology, the first step is to take input images from the

TABLE 8.1

Performance of Proposed System

S. No.	Input Image	Segmented Image	Number of Iterations	Segmentation Time	Area of the Neoplasm
1			230	6.67	0.0105
2			280	7.12	0.0845
3			180	9.12	0.0586
4			75	4.65	0.0326

database, adding white Gaussian noise to this additive. The resultant image is passed through a de-noising filter called a BF. An MRI picture was segmented using active contour segmentation. The initial contour selection and the number of iterations significantly influence accurately identifying cancer from a digital image.

When used for brain tumor segmentation in MRI images, the suggested method would be beneficial to physicians, clinical specialists, and the analyst's location in addition to the analyst himself. The process may also be extended to various other clinical imaging modalities to get precise clinical data as fast and efficiently as possible. It is also possible to segment the pictures of the check-up for simple inspection

TABLE 8.2
Performance of Proposed Algorithm at Different Iterations

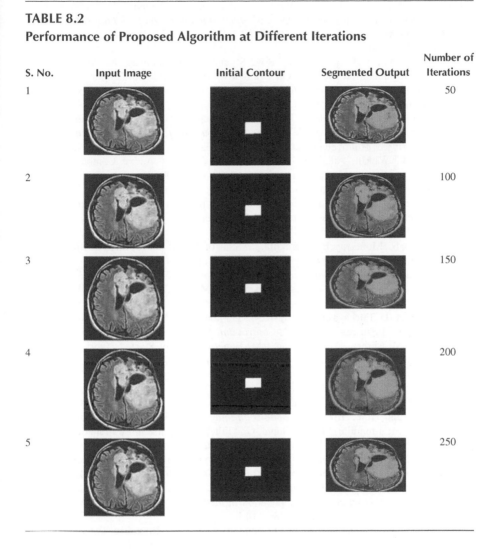

S. No.	Input Image	Initial Contour	Segmented Output	Number of Iterations
1				50
2				100
3				150
4				200
5				250

and identification of lesions and blood clotting, abnormal outgrowth, and cancer cells in cysts, tumors, tiny aneurysms, and inflations.

REFERENCES

[1] Kostas, P. "Non-Rigid Registration of Medical Images Using an Automated Method," *International Journal of Medical and Health Sciences*, 1, 7, 477–479, 2007.

[2] J. Jaya, K. Thanushkodi, "Exploration on selection of medical images employing New Transformation Techniques," *IJCSI International Journal of Computer Science Issues*, 7, 3, 4, 2010.

[3] K Gurucharan, S S Kiran, K Babburu and L Vadda, "Computer Vision-Based Fruit Recognition and Classification System," *International Journal of Future Generation Communication and Networking (IJFGCN)*, ISSN: 2233–7857(Print); 2207–9645(Online), 2020, 13, 3, 1–14.

[4] Nag, S, Maitra, I, Sudipta, M, Roy, S, and Bandyopadhyay, K, "A Review of Image Segmentation Methods On Brain MRI for Detection of Tumor and Related Abnormalities," *International Journal of Advanced Research in Computer Science and Software Engineering*, 4, 5, 1073–1095, 2014.

[5] Bandyopadhyay S, and Paul T.," Automatic Segmentation of Brain Tumour from Multiple Images of Brain MRI," *International Journal of Application or Innovation in Engineering & Management (IJAIEM)*, 2, 1, 2013.

[6] Kass, M., Witkin, A.P., and Terzopoulos, D. "Snakes: Active Contour Models," *International Journal of Computer Vision*, 1, 321–331, 2004.

[7] Menze, H., Jakab, A., Bauer, S., Kalpathy-Cramer, J., Farahani, K., Kirby, J., Burren, Y., Porz, N., Slotboom, J., Wiest, R., et al., "The Multimodal Brain Tumor Image Segmentation Benchmark (Brats)," *IEEE Transactions on Medical Imaging*, 34, 1993–2024, 2015.

[8] Gordillo, N., Montseny, E., Sobrevilla, P. "State of the Art Survey on MRI Brain Tumor Segmentation," *Magnetic Resonance Imaging*, 31, 8, 1426–1438, 2013.

[9] Nabizadeh, N. *"Automated Brain Lesion Detection and Segmentation Using Magnetic Resonance Images."* Ph.D. Thesis, University of Miami, Coral Gables, FL, USA, April 2015.

[10] R. Biswas, D. Purkayastha and S. Roy, "De-noising of MRI Images Using Curvelet Transform. Advances in Systems", *Control and Automation*, 575–583, 2017.

[11] K. Das, M. Maitra, P. Sharma, and M. Banerjee, "Early Started Hybrid Denoising Technique for Medical Images," *Recent Trends in Signal and Image Processing*, 131–140, 2018.

[12] Lysaker, M., Lundervold, A., and Tai, X.-C., "Noise Removal Using Fourth-Order Partial Differential Equation with Applications to Medical Magnetic Resonance Images in Space and Time," *IEEE Transactions on Image Process*, 12, 12, 1579–1590, 2003.

[13] Arora, S., Hanmandlu, M., and Gupta, G., "Filtering impulse noise in medical images using information sets," *Pattern Recognition Letters*, 2018.

[14] Bai, J., Song, S., Fan, T. and Jiao, L., "Medical image denoising based on sparse dictionary learning and cluster ensemble," *Soft Computing*, 22, 5, 1467–1473, 2018.

[15] Lee, Y., "Improved total-variation noise-reduction technique with gradient method using iteration counter and its application in medical diagnostic chest and abdominal X-ray imaging," *Optik*, 170, 475–483, 2018.

[16] Caldairou, B., Passat, N., Habas, P., Studholme, C., and Rousseau, F., "A non-local fuzzy segmentation method: Application to brain MRI". *Pattern Recognition*, 44, 9, 1916–1927, 2011.

[17] Yang, J., Fan, J., Ai, D., Wang, X., Zheng, Y., Tang, S., and Wang, Y., "Local statistics and non-local mean filter for speckle noise reduction in medical ultrasound image," *Neurocomputing*, 195, 88–95, 2016.

[18] Gai, S., Zhang, B., Yang, C., and Yu, L., "Speckle noise reduction in medical ultrasound image using monogenic wavelet and Laplace mixture distribution," *Digital Signal Processing* 72, 192–207, 2018.

[19] Sudeep, P., Palanisamy, P., Rajan, J., Baradaran, H., Saba, L., Gupta, A., and Suri, J., "Speckle reduction in medical ultrasound images using an unbiased non-local means method," *Biomedical Signal Processing and Control*, 28, 1–8, 2016.

[20] Zhang, Y., Tian, X., and Ren, P., "An adaptive bilateral filter based framework for image de-noising," *Neurocomputing*, 140, 299–316, 2014.

[21] Qi, M., Zhou, Z., Liu, J., Cao, J., Wang, H., Yan, A., Wu, D., Zhang, H., and Tang, L., "Image De-noising Algorithm via Spatially Adaptive Bilateral Filtering." *Advanced Materials Research*, 760–762, 1515–1518, 2013.

[22] Zhang, M., and Gunturk, B., "Multiresolution Bilateral Filtering for Image De-noising." *IEEE Transactions on Image Process*, 17, 12, 2324–2333, 2008.

[23] Ramananda, S., Lehmann, B., and Kraus, D., "Optimal parameters for bilateral filtering and SAS image de-noising," *2011 International Symposium on Ocean Electronics*, 2011.

[24] Shi, H. "Determination of bilateral filter coefficients based on particle swarm optimization," *2012 5th International Congress on Image and Signal Processing*, 302–306, 2012.

[25] Farooq, M. "Application of genetic algorithm & morphological operations for image segmentation," *International Journal of Advanced Research in Computer and Communication Engineering*, 4, 3, 2015.

[26] D. Bhattacharyya, K Taihoon, "Brain Tumor Detection Using MRI Image Analysis," *Communications in Computer and Information Science*, 151, 2011, 307–314.

[27] M. H. Chowdhury, W. D. Little., "Image thresholding techniques" *IEEE Pacific Rim Conference on Communications, Computers And Signal Processing, Proceedings 17–19 May 1995*, 1995, 585–589.

[28] P. K. Sahoo, S. Soltani, A. K. C. Wong, and Y. C. Chen, "A Survey of thresholding techniques," *Computer Vision Graphics Image Process (CVGIP)*, 41, 1988, 233–260.

[29] M. Sezgin, B. Sankar, "Survey over image thresholding techniques and Quantitative performance evaluation," *Journal of Electron Imaging*, 13, 1, 2004, 146–165.

[30] S. Xavierarockiaraj, K. Nithya, R. Maruni Devi, "Brain tumor Detection using Modified Histogram thresholding Quadrant approach," *Journal of Computer Applications*, 5, 1, 2012, 21–25.

[31] P. L. Lions, J. M. Morel, T. Coll, "Image selective smoothing and edge detection by nonlinear diffusion." *SIAM Journal of Numerical Analysis*, 29, 1, 1992, pp.182–193.

[32] R. C. Gonzalez, E. W. Richard, *"Digital Image Processing,* Pearson Education. 3rd Edition. 2007

[33] N. R. Pal, S. K. Pal, "A Review on Image Segmentation Techniques," *Pattern Recognition*, 26, 9, 1993, 1277–1294.

[34] W. X. Kang, Q. Q. Yang, R. R. Liang, "The Comparative Research on Image Segmentation Algorithms," *IEEE Conference on ETCS*, 2009, 703–707

[35] R. C. Gonzalez, E. W. Richard, *"Digital Image Processing,* Pearson Education. 3rd Edition. 2007

[36] S. Varshney, N. Rajpal, R. Purwar, "Comparative Study of Image Segmentation Techniques and Object Matching using Segmentation," *Proceeding of International Conference on Methods and Models in Computer Science*, 2009, 1–6.

[37] H. Zhang, J. E. Fritts, S. A. Goldman, "Image Segmentation Evaluation: A Survey of unsupervised methods, " *computer vision and image understanding*, 2008, 260–280.

[38] L. Pharm Dzung, X. Chenyang, J. L. Prince, "A Survey of Current Methods in Medical Image Segmentation," *Technical Report JHU / ECE 99-01, Department of Electrical and Computer Engineering*, 1998.

[39] X. Jiang, R. Zhang and S. Nie, "Image Segmentation Based on PDEs Model: A Survey," *2009 3rd International Conference on Bioinformatics and Biomedical Engineering*, 2009, 1–4.

[40] Damodaran, S. and R. Dhanasekaran, "Segmentation of cerebrospinal fluid and internal brain nuclei in brain magnetic resonance images," *International Review on Computers and Software,* Vol.8, 5, 2013, 1063–1071.

[41] Chan, T. F., and L. A. Vese. "Active contours without edges." *IEEE Transactions on Image Processing: A Publication of the IEEE Signal Processing Society*, 10, 266–277, 2001.

[42] T. Zhang, Y. Xia and D. D. Feng, "A deformable segmentation algorithm for brain M.R. images," *2012 Annual International Conference of the IEEE Engineering in Medicine and Biology Society*, 2012, 3215–3218

[43] D. Mumford and J. Shah, "Optimal approximations by piecewise smooth functions and associated variation problems," *Communications on Pure and Applied Mathematics*, 42, 577–685, 1989.

[44] C. Tomasi and R. Manduchi, "Bilateral Filtering for Gray and Color Images," *Sixth International Conference on Computer Vision (IEEE Cat. No.98CH36271)*, 1998, 839–84.

[45] Rami Cohen, "The Chan-Vese Algorithm," *Computer Vision and Pattern Recognition*, arXiv:1107.2782.

[46] K Babburu, K Gurucharan, S S Kiran, P Srujana, *PAPR Reduction Techniques, and Image Quality Assessment in Image-Based MMS VLC System*, I.T. in Industry, 9, 2, 2021.

9 Machine Learning Techniques Applied to Extract Objects from Images
Research Issues Challenges and a Case Study

*Reena Thakur, Pradnya S. Borkar,
and Parul Bhanarkar*
Jhulelal Institute of Technology, Nagpur

Prashant Panse
Medi-Caps University

CONTENTS

DOI: 10.1201/9781003248750-9

9.1 INTRODUCTION

This world includes a million types of living things. How can anybody, then, differentiate between them? Thus, it is necessary to distinguish between different things, color, smell and day of life for all living beings. In order for any action to be carried out by any living being, it is important to have expertise to distinguish between different inputs that they obtain through their sensory organs from the external world. In each day-to-day task, beginning in the morning, humans easily distinguish one object from another. The object here for this work is the factual entity which can be seen and observed. It can include simple objects including things like your car and TV. There are various objects that can be seen in the images and videos that are captured.

The object detection as shown in Figure 9.1 is the computer vision task similar to image classification, object recognition, object segmentation or semantic segmentation. With the recent advancement in the deep learning algorithms in computer vision models, mentioned tasks have gained importance enabling improvement in performance and enabling real-time use cases.

The image classification as shown in Figure 9.2 actually assigns class label to the objects in the class and then identifies the class of objects. Image classification works using the image features. Object localization also is one of the computer image pre-processing visions tasks wherein a springing box is strained about the object inside a given image, hence identifying the location of the object of interest.

The dimension of an object is a part of observance of the appearance of the object. Out of the many uses of object dimension, the dimensions could be used in extensive surveys and estimations. Complex levels of image processing may be needed in these cases to find the precise dimension. Some of the generalized tasks include the following:

1. Predicting the minimum and maximum dimensions of parts and products
2. Estimating precise dimensions about the roundness and angles of the metal parts for accurate assembly.

The task of object detection and dimension prediction is related to transforming simple, unreliable low-level measurements/dimension values to complex, high-level descriptions/heuristics of dimensions. The object detection as well as object

FIGURE 9.1   Individual object detection (lamp, sofa table, table, flower pots).

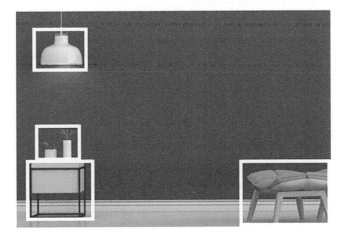

FIGURE 9.2   Image classification.

recognition has numerous applications in the field of computer vision and has wide scope of research and study. Applications to name a few are self-driving cars, surveillance system and image/object extraction. The object dimension prediction forms the next stage of the object detection and recognition.

To know size of an object, image segmentation can help to segment the required area from the image making it one step easier to identify its shape and size. The image segmentation and object detection follow the implementation pattern from convolutional neural networks (CNNs), region-based convolutional neural network (RCNN), fast RCNN, Mask RCNN and You Only Look Once (YOLO) algorithms.

9.2 MOTIVATION

Object detection, object dimension reduction and object recognition and tracking using machine learning and computer vision have gained greater importance during the past few years. Motivation for conducting research in this area will help identify problems in the field and also to answer to certain uncertainties. In this regard, it is also important that the result of the research is in some manner benefit the society or human race. As one can note, the tremendous growth in computer vision under the category of machine learning and various applications of machine learning methods can be witnessed to solve the computer vision tasks. Certain applications of the same include the object detection and tracking, identifying persons smiling, image-based web searches and 3D image reconstruction to name a few. The computer vision technology will in future act as the solutions provider for solving various real-world problems. The modern computer vision uses deep learning to develop algorithms that allow computers to process the visual input data and then extract some useful insights. The above technology is nowadays worth adopting in the commercial and non-commercial fields to strengthen various operations.

9.3 CHALLENGES AND ISSUES

Object detection methods have the ability to detect specific as well as multiple objects at once in an image or video frame. Challenges arise when we try to track the objects in consecutive frames of videos. The challenges include object motion, noisy image, distorted videos, irregular object shape, occlusion of object and constant changes in position. The object recognition algorithms work over output of object detection methods and uses detected objects/interested object features from the input image or videos to classify objects into certain category, hence, recognizing the exact object. There are certain challenges including the size on the sample space which is used as input, uncertainty involved in the input data, ambiguous information about objects of interest. The object dimension prediction challenges include the identify the bounding box in the stable or moving objects including varied and complex objects, finding the visual relationship and finding the dimension of the objects identified. The issues that need to be handled for the successful object detection are discussed in detail as shown in Figure 9.3.

9.4 ARCHITECTURE

To get more accurate segmentation, the entire image is to be segmented to extract an object from an image in terms of saving global features. Furthermore, extract the projected object from the image by integrating the segments that are inside the area drawn previously by us as shown in Figure 9.4.

a. **Pre-processing of Images:** The main purpose of pre-processing is enhancing the image features by reducing or removing the unwanted data or defects and also boosting the critical image qualities, so that

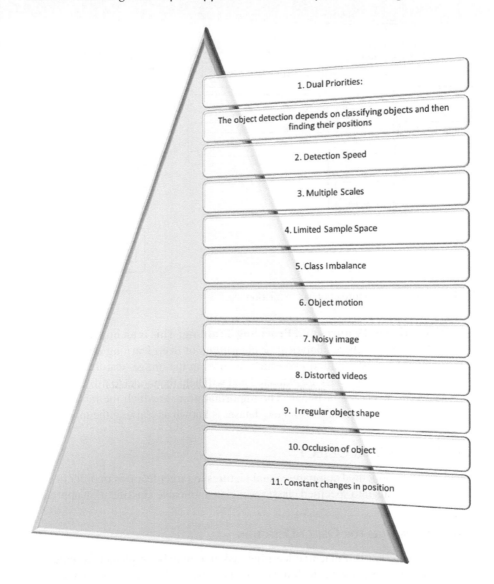

FIGURE 9.3 Issues.

our computer vision architectures are able to work with good data. Pre-processing is a word that refers to image processing at fundamental abstraction level, where the input and output both are intensity pixels. Pre-processing improves the image features by removing unwanted distortions and thereby improving certain visual aspects in the image which are beneficial for further operations.

b. **Detecting Object:** The process of identifying an object is known as "detection," and it comprises segmenting image and establishing position of an object of interest.

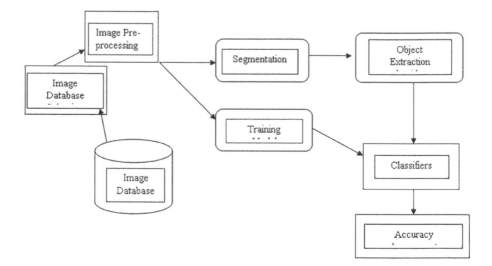

FIGURE 9.4 Architecture of object extraction.

c. **Extraction of Features and Providing Training:** This is an important step in the process where analytical or algorithms of deep learning applied for finding the most required interesting features in the image; features unique to a specific image class are extracted and will help the model to differentiate between the classes of image. The important process which the model follows to learn the features from the dataset is known as training the model.

9.5 CLASSIFYING OBJECT

The classification method compares visual features to target features in order to classify observed items into specified classes using an applicable classification approach.

9.5.1 TECHNIQUES FOR OBJECT DETECTION

The object detection is a machine learning task that may be implemented either by traditional image processing techniques or the advanced deep learning methods. The most recent development in the deep convolutional network and the GPUs that have increased computing power can be exclusively utilized for object detection. The advancements in the object detection algorithm have made it possible to achieve high accuracy in detection. The object detectors called one-stage and two-stage object detectors are represented in Figure 9.5:

The above-presented object detection algorithms basically perform the following tasks:

1. Determining the objects from the input image or video.
2. The object recognition or classification and later find the size of the bounding box representing the object accurately.

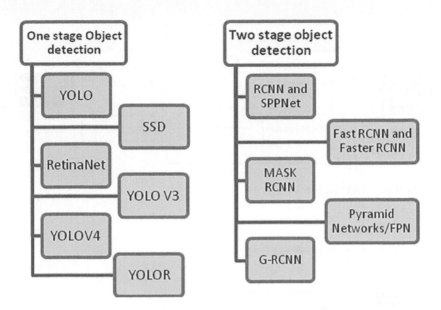

FIGURE 9.5 Object detection algorithms.

The two-stage detectors perform object region proposals and then perform the feature-extraction–based object classification. The one-stage detection methods can work without finding the region proposal work. The methods usually predict the bounding box over the input image or video.

9.5.2 REGION PROPOSAL NETWORKS

The region proposal networks form the backbone for an object detection process in the two-stage detectors. The reason behind using the RPN is the identification of the available objects in the input image helping toward real-time object detection.

9.5.2.1 One-Stage Object Detection Algorithms
a. YOLO
YOLO is the You Only Look Once object detection algorithm for fast object detection which has benefits like speed and accuracy as compared to other algorithms.

The object detection problem in YOLO can be considered as the regression problem to detect the objects in the real-time input through bounding box. The noted applications include detection of traffic signals, animals and people. The YOLO bounding box regression example is shown in Figure 9.6:

b. Bounding Box Regression
The bounding box method is used to identify possible location of an object in the image and outline the object to highlight it as shown in Figure 9.7. Each created bounding box consists of width, height, class and

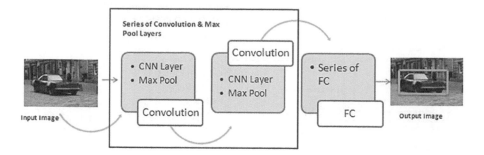

FIGURE 9.6   YOLO architecture.

FIGURE 9.7   Bounding box regression.

bounding box center. The center is defined as (cx, cy), width of box by width (wb), height of box by height (bh) and class of object by c.

c. YOLO Applications

- **Autonomous Driving [1]:**

    The YOLO deep learning algorithm with the bounding box regression can be applied for detection of obstacles and objects like people and vehicles at real time for the autonomous driving vehicles to avoid accidents happening.

- **Surveillance Security Systems [2]:**

    In the security domain, YOLO object detection can be applied to check/enforce the security in a restricted area. The intelligent video surveillance developed using YOLO is now able to perform some required image processing tasks which help in human behavior prediction and safety.

- **Animal Detection in Wildlife [3]:**

    The YOLO object detection can be helpful to the wildlife departments for real-time recording and observation of the wild life. The challenge here is to detect the animals with their names and their classification. The model can be designed and tested using pre-trained datasets.

- **Human Emotions Recognition [4]:**

    Emotion analysis can be done by the extraction of facial features which further helps identify and examine a person in stress or exhilaration or depression etc. The face thermal images can be analyzed using the YOLO deep learning algorithms.

- **Real-Time Traffic Counting [5]:**

    The traffic information can be utilized in efficient management of the transportation system. The real-time vehicle flow counting can be automated with the help of YOLO object detection algorithm. The bounding box regression is applied to perform this task effectively and avoid false detection.

d. SSD (Single-Shot Detectors) [6]:

This detector is the fast SSD used for object detection tasks. This model of object detection works on the convolutional network which generates bounding boxes of fixed sizes.

SSD Applications:

- **SSD for Autonomous Vehicles [7]:**

    In Industry 4.0, the autonomous vehicles form a smart device that can be used to perform certain kinds of navigation tasks or even shipping and transfer of goods.

- **Feature Aggregation and Enhancement [8]:**

    The enhancement of the shallow and deep feature maps is crucial for efficient object detection. The feature aggregation and enhancement can be integrated with the SSD in order to achieve higher accuracy.

- **Real-Time Detection for Camera for Camera Sensing System [9]:**

    The camera sensing system is an important device for object detection with the help of single-shot multi-box detector. This method is helpful in reducing the overfitting problem in object detection.

e. RetinaNet:

The one-stage object detector called RetinaNet is the one applicable for small objects detection. The RetinaNet till now has been applied for satellite and areal imagery. It is designed over the combination of features from the focal loss method and featured pyramid network.

RetinaNet Applications:

- **3D RetinaNet for Underwater Moving Object Detection [10]:**

    The 3D RetinaNet can be applied for the challenging imaging conditions like complex background and low visibility in the water. The marine life science study for finding information can use this model for detection and fish tracking.

- **Visual Inspection of Flexible Materials [11]:**

  The amount of background information in visual inspections of flexible materials, such as textile surfaces, is excessive, resulting in the dilemma that the main qualities that must be retrieved are not readily apparent. The focal error rate is utilized to build RetinaNet network and faster filtering of background noise/information, allowing input image feature information to be acquired at a faster rate, despite the fact that input considered has defective surface.

- **Detection of Hand Movement [12]:**

  Multi-scale context information extraction can be useful for the intelligent homecare systems. The hand detection could be performed using the atrous convolution module along with RetinaNet for improving the detection performance.

f. YOLOV3 and YOLOV4:

The YOLOV3 is better and faster than the SSD and RetinaNet and also does not require retraining. The YOLOV3 is applied to the input image at multiple locations and multiple scales. The regions are considered for performing detection based on the high scores. The bounding prediction is done after dividing the image into several regions. These bounding boxes have weighted probabilities. The normalization and anchor methods can be used for predicting the bounding box. The YOLOV4 is an extension of YOLOV3 applied for deep-learning–based real-time detection of interested objects. The YOLOV4 uses the pyramid pooling and the path aggregation networking order to achieve speed and accuracy.

YOLOV3 and YOLOV4 Applications:

- **Multiple Object Detection [13]:**

  YOLOV3 and YOLOV4 object detection algorithms can be successful for smart surveillance and traffic control applications. The models of object detection are also used for feature extraction from scenarios including movies and computer games.

- **Advanced Traffic Sign Recognition [15]:**

  For the effective traffic sign detection and recognition, various annotated training data are required. The CNN-based visual system could be used to determine the quality of data. The Generative Adversarial Networks (GANs) may be used to improve the databases of signs, adding more accurate and diverse training dataset.

g. YOLOR (You Only Learn Once Representation)

The YOLOR model used for object detection is an algorithm which works on the implicit and explicit knowledge together. The unified network is used here to produce the unified representation. The architecture has three processes to do object detection including the kernel space alignment and prediction refinement through CNN.

YOLOR Applications:

- **Multi-Model Remote Sensing [14]:**

The recently developed YOLOR can be used in order to improve real-time detection performance by applying fusion of better multiple remote-sensing technologies.

9.5.2.2 Two-Stage Object Detection Algorithms

a. RCNN and Spatial Pyramid Pooling CNN (SPPNet)

The SPPNet is the popular image classification and object detection method which works on the variable size input image. It helped the bounding box prediction speed as compared to other methods. The RCNN method generates the category-independent RPNs and extracts the fixed length feature vectors from CNN models. They use the superpixel-based selective search.

RCNN and SPPNet Applications:

- **Traffic Classification [16]:**

    The SPPNet could be used by applying the method of max-pooling layer that can be utilized for analysis of network traffic and the variable length dataset.

- **Intelligent Transportation Systems [17]:**

    Traffic monitoring, security systems, surveillance systems are the application areas where the RCNN has shown good performance.

b. Fast RCNN and the Faster RCNN:

The fast RCNN was designed to improve or reduce the drawbacks from the RCNN and SPPNet.

Fast RCNN and Faster RCNN Applications:

- **Object Detection for Optimal Remote-Sensing Images [18]:**

    The object detection finds its applications in many aerials and satellite image analysis. The images captured by the remote-sensing image devices may contain complex backgrounds and many other complexities. To deal with these, the faster RCNN along with feature pyramid networks and deformable convolution network could be used which also increases the detection performance.

- **Event Analysis for Vehicle Classification [19]:**

    The vehicle classification and counting has formed an integral part in the traffic management to control and make analysis of the traffic flow. The fast RCNN can be utilized for the successful real-time vehicle classification by extracting the features from the event video streaming.

- **Classification and Recognition over Documents [20]:**

    The limitation of the current OCR technology is that it can only process simple digital documents. The faster CNN can be utilized for processing of the complex documents involving non-textual contents like noise, signatures and forms.

c. Mask RCNN:

Oriented on the concept of feature pyramid network, this mask RCNN is the effective object detection as well as segmentation method. This model helps to find out the bounding boxes and the segmentation masks for all the instances of objects appearing in the input image

Mask RCNN Applications:

- Target detection [21]:

  In many robotic applications, one of the major challenges is implementing the target detection in order to avoid collision at workplace. This also has impact on the robot object recognition time and target capture.

- **Human Body Attitude Detection [22]:**

  The mask RCNN provides higher accuracy for detection models that can be utilized for the multi-person attitude detection. The detection speed can be further improved with the MobileNet-based mask RCNN method.

d. Pyramid Networks/FPN:

The feature pyramid networks are simple frameworks under the CNN. The Feature Pyramid Network (FPN) has the capability to integrate the image features that are low resolution with the high-resolution features, which makes detection easier.

FPN Applications:

- **Action Recognition [23]:**

  The two-pathway networks along with the feature pyramid networks can be utilized for implementing the action recognition in many applications. The network also handles the fusion of the temporal and spatial image features to assist the action recognition better.

- **Medical Image Segmentation [24]:**

  The detection applied over the medical images is vital for the analysis after medical diagnosis. The FPNs can be utilized for performing the accurate and automatic segmentation for medical image.

e. G-RCNN [26]

The graph recognition CNN is the extension of the RCNN. It utilizes the serious component of RCNN model, that is, the recurrent layer of convolution through which recurrent types of connections among given neurons can be incorporated creating the gated recurrent convolutional layer.

G-RCNN Applications:

- **Autonomous Tool Construction [25]:**

  In the area of robotics, the significant challenge is bringing automation in the tool construction which involves the process of reconstruction of a new tool based on a reference tool.

9.6   APPLICATIONS OF MACHINE LEARNING

9.6.1   Machine Learning in Computer Vision

Computer vision tasks are tasks that replicate the human vision tasks. The human vision tasks include detecting the objects and then labeling them. This is done by a very complex organic process involving the eyes and the cortex. The human can do this using the stored and learnt models of different objects from the real world and the understanding of the concept. This also includes the experiences of the interactions to the environment.

9.6.2 OBJECT DETECTION AND DIMENSIONALITY PREDICTION USING MACHINE LEARNING AND DEEP LEARNING

Ever wondered while flying over the ocean and spotting several ships of various sizes and wondering what may be the size and shape of each of them. Object dimensionality can be found out after the object in the image is detected. The various machine and deep learning methods can be devised in order to analyze dimensions of a particular object. The object recognition algorithms work over output of object detection methods and use detected objects/interested object features from the input image or videos to classify objects into certain category, hence recognizing exact object.

9.7 IMAGE CLASSIFICATION

Object classification is a fairly straightforward exercise for human, however, which shows significant problem for recent machines; hence, image classification has in recent times become a critical area of research. The process of classifying an image is done by categorizing images into one class out of the several predetermined categories. Single image can be classified into an endless number of categories. The automation of the classification process with computer vision would be definitely helpful.

Advances in the field of autonomous driving are also an excellent example of how picture categorization may be used nowadays.

9.8 CASE STUDIES [1]

Some examples are given below that show how the image gets processed completely.
The steps for image pre-processing are shown in Figure 9.8.

9.8.1 READING IMAGE

At this phase, we save the complete path to the set of images under consideration in a particular variable and then a function is written in order to load image files into arrays structure, so that computers are able to manage them.

9.8.2 RESIZE IMAGE

Because the sizes of some photos acquired by a specific camera and provided to our artificial intelligence algorithm differ, we need to resize every image fed to our artificial intelligence algorithms to establish a base size.

9.8.3 DATA AUGMENTATION

Data augmentation is a technique for generating fresh "data" with various orientations. This has two advantages: first, it allows you to make "additional data" from inadequate data, and second, it prevents overfitting.

FIGURE 9.8   Steps for image pre-processing.

9.8.3.1   Techniques-Data Augmentation

1. Grayscaling
2. Flip or reflection
3. Gaussian blurring
4. Histogram equalization
5. Rotation
6. Translation

9.9   IMAGE CLASSIFICATION TECHNIQUES

This section covers statistical classifiers under machine learning which includes decision trees and the SVMs, that is, support vector machines, and then describes CNNs a deep learning technique.

The findings of the classification task can be used to discriminate lymphoblastic type of leukemia cells which are different from non-lymphoblastic cells that have

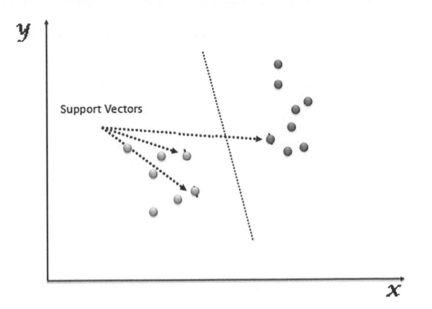

FIGURE 9.9   SVM classifier.

been supplied to enhance their performance analysis. CNN is a deep learning technique to extract the features, which will also be explored as one of our classifiers.

Different classifiers are then applied to this feature extractor in order to classify the data.

9.9.1   SVM Models

SVM model is a type of supervised machine learning approach which may be applied to classification and regression types of tasks. A linear boundary method is used to distinguish the classes for classification as shown in Figure 9.9.

In a given high-dimensional space, it generates the hyperplane or a continued sequence of hyperplanes, with the hyperplane having largest distance to nearest training sample data point of each considered class achieving stable separation between the two considered classes. An algorithm's true power is determined by the kernel function that is used.

The most often used kernel models are as follows:

- Linear kernel model
- Gaussian kernel model
- Polynomial kernel model

9.9.2   Decision Tree Models

Decision tree models are supervised machine learning algorithms which use only the tree data structure and a few if/else statements on the features chosen.

The rejection as well as acceptance of class labels at each intermediate stage or level is possible with a hierarchical rule-based system, decision trees as shown in Figure 9.10.

9.9.3   k-Nearest Neighbor Model

The method of k-nearest neighbor is the basic machine learning algorithm. The mentioned technique classifies unidentified image data points by way of identifying utmost frequent class among all the k-closest considered samples based solely over the distance given between the feature vectors as shown in Figure 9.11.

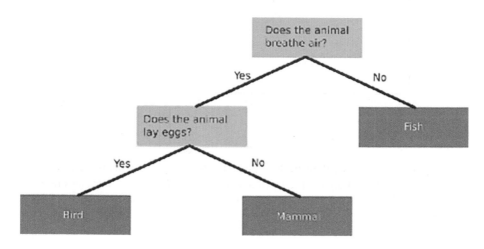

FIGURE 9.10   Decision tree example.

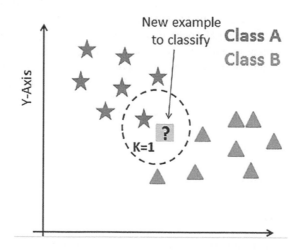

FIGURE 9.11   k-Nearest neighbor classifier.

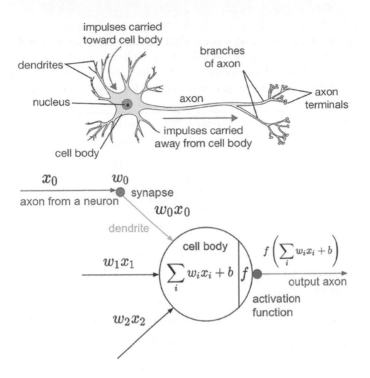

FIGURE 9.12   (a) Block diagram of ANN. (b) ANN layers.

9.9.4 ARTIFICIAL NEURAL NETWORK (ANN) MODELS

ANN models shown in Figure 9.12a are the different statistical learning models/ algorithms which are inspired by biological neural networks. They are utilized for different tasks and areas, from basic categorization to computer vision as well as speech recognition tasks.

Numerical values are stored as weights over the node-to-node connections. The network can eventually utilize the intended functions by adjusting the generated values in a manner systematic as shown in Figure 9.12b.

Each hidden layer feature detector can be an individual feature detector that can detect increasingly complex patterns in the given data as it passes through the developed network. If this network is assigned the recognition of face then the first hidden layer can act as a type of line detector, the second hidden layer takes these lines as input variable and then combines them to form parts like nose, the third hidden layer might take the nose and matches it with eye and so on, until entire face image is built using these features [28,29].

9.9.5 CNN MODEL

CNN models are deep learning techniques and again a type of ANN with a unique architecture. CNNs have achieved up-to-the-minute performance in the field of computer vision challenges by utilization of some of the visual cortex's attributes. CNNs are made of two basic components: convolution and pooling layer. Furthermore, nearly limitless conducts are available for assembling all layers for a particular issue under consideration. Convolutional and pooling layers are simple to comprehend as seen in a CNN. The most complex aspect of deploying CNNs is determining the appropriate model topologies shown in Figure 9.13.

Classifier	Accuracy (%)	Precision	Recall	ROC
SVM	85.68	0.86	0.87	0.86
Decision trees	84.61	0.85	0.84	0.82
K-nearest neighbor (KNN)	86.32	0.86	0.86	0.88
ANN (for 100 epochs)	83.10	0.88	0.87	0.88
CNN (for 300 epochs)	91.11	0.93	0.89	0.97

Some more examples with tools are depicted in Table 9.1.

9.10 CONCLUSION

Thus, in this chapter, various techniques of machine learning algorithms used to extract objects from images have been described as well as case studies were also discussed. First, the related works as well as background knowledge were introduced in detail. In addition, in what way machine learning algorithms are applied in the monarchy of extracting objects from images, from the perspective of optimization techniques, resource management and security have been reviewed. Finally, case studies are discussed, which are worthwhile to be pursued by the researchers in the future.

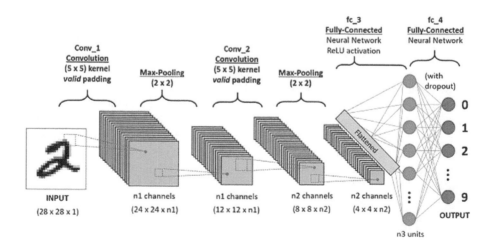

FIGURE 9.13   Steps follows while applying CNN.

TABLE 9.1
Examples of Tools

Name of Tool	Image	Resource	Function	Outcome
Brandfolder		https://brandfolder. com/workbench/extract-text- from-image	You may extract text from any image with the text extractor. You can submit an image or a PDF document, and the application will extract text from it. You can copy to your clipboard with a single click once the file has been extracted	Society 4.0 Cyberspace 0006 Person access, retrieve and analyze the information Clouds Physical space in library Physical space for banking Robots under control of human Physical space
ImageData Extractor 2.0		http://www.imagedataextractor.org /results/f85665a1-695b-4a74- b6af-1a4443b271c9	ImageDataExtractor is able to extract data from images where individual particles in an image can be discerned and separated by a human	  Particle Instance Segmentation Map

(*Continued*)

TABLE 9.1 (*Continued*)
Examples of Tools

Name of Tool	Image	Resource	Function	Outcome

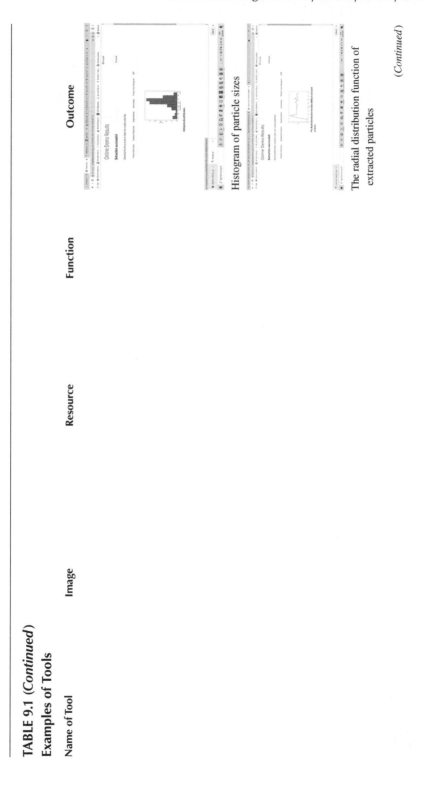

Histogram of particle sizes

The radial distribution function of extracted particles

(*Continued*)

TABLE 9.1 (*Continued*)
Examples of Tools

Name of Tool	Image	Resource	Function	Outcome
Image analysis tool		https://compress-or-die.com/analyze-process	This is a free online analysis tool applied over images that generates a lot of information found in the input image	
Wordcloud (cloud creators are used to highlight popular words and phrases)	Abstract of this chapter	https://monkeylearn.com/word-cloud/rest.lt		

REFERENCES

1. https://iq.opengenus.org/basics-of-machine-learning-image-classification-techniques/.
2. A. Sarda, S. Dixit and A. Bhan, "Object Detection for Autonomous Driving using YOLO [You Only Look Once] algorithm," *2021 Third International Conference on Intelligent Communication Technologies and Virtual Mobile Networks (ICICV)*, 2021, pp. 1370–1374, doi: 10.1109/ICICV50876.2021.9388577.
3. H. H. Nguyen, T. N. Ta, N. C. Nguyen, V. T. Bui, H. M. Pham and D. M. Nguyen, "YOLO Based Real-Time Human Detection for Smart Video Surveillance at the Edge," *2020 IEEE Eighth International Conference on Communications and Electronics (ICCE)*, 2021, pp. 439–444, doi: 10.1109/ICCE48956.2021.9352144.
4. B. K. Reddy, S. Bano, G. G. Reddy, R. Kommineni and P. Y. Reddy, "Convolutional Network based Animal Recognition using YOLO and Darknet," *2021 6th International Conference on Inventive Computation Technologies (ICICT)*, 2021, pp. 1198–1203, doi: 10.1109/ICICT50816.2021.9358620.
5. Chaitanya, S. Sarath, Malavika, Prasanna and Karthik, "Human Emotions Recognition from Thermal Images using Yolo Algorithm," *2020 International Conference on Communication and Signal Processing (ICCSP)*, 2020, pp. 1139–1142, doi: 10.1109/ICCSP48568.2020.9182148.
6. J. Lin and M. Sun, "A YOLO-Based Traffic Counting System," *2018 Conference on Technologies and Applications of Artificial Intelligence (TAAI)*, 2018, pp. 82–85, doi: 10.1109/TAAI.2018.00027.
7. Liu W. et al. SSD: Single Shot MultiBox Detector. In: Leibe B., Matas J., Sebe N., Welling M. (eds.), *Computer Vision – ECCV 2016. ECCV 2016. Lecture Notes in Computer Science*, vol. 9905, 2016. Springer, Cham. doi: 10.1007/978-3-319-46448-0_2.
8. GENGEÇ, N., EKER, O., Cevikalp, H., Yazici, A., and Yavuz, H. Visual object detection for autonomous transport vehicles in smart factories. *Turkish Journal of Electrical Engineering & Computer Sciences*, 2021, 29, 2101–2115. doi: 10.3906/elk-2008-62.
9. W. Li and G. Liu, "A Single-Shot Object Detector with Feature Aggregation and Enhancement," *2019 IEEE International Conference on Image Processing (ICIP)*, 2019, pp. 3910–3914, doi: 10.1109/ICIP.2019.8803543.
10. B. Ding, L. Gu and X. Zhu, "Real-Time Detection for Camera Sensing System: Adaptive Cascade Single-Shot Detector," *2019 14th IEEE Conference on Industrial Electronics and Applications (ICIEA)*, 2019, pp. 2154–2160, doi: 10.1109/ICIEA.2019.8834260.
11. Shen and C. Nguyen, "Temporal 3D RetinaNet for fish detection," *2020 Digital Image Computing: Techniques and Applications (DICTA)*, 2020, pp. 1–5, doi: 10.1109/DICTA51227.2020.9363372.
12. W. Wu, L. Wu, J. Li, S. Wang, G. Zheng and X. He, "RetinaNet-Based Visual Inspection of Flexible Materials," *2019 IEEE International Conference on Smart Internet of Things (SmartIoT)*, 2019, pp. 432–435, doi: 10.1109/SmartIoT.2019.00077.
13. Q.-V. Hoang, T.-H. Le and S.-C. Huang, "An Improvement of RetinaNet for Hand Detection in Intelligent Homecare Systems," *2020 IEEE International Conference on Consumer Electronics – Taiwan (ICCE-Taiwan)*, 2020, pp. 1–2, doi: 10.1109/ICCE-Taiwan49838.2020.9258335.
14. C. Kumar B. R. Punitha and Mohana, "YOLOv3 and YOLOv4: Multiple Object Detection for Surveillance Applications," *2020 Third International Conference on Smart Systems and Inventive Technology (ICSSIT)*, 2020, pp. 1316–1321, doi: 10.1109/ICSSIT48917.2020.9214094.
15. M. Sharma et al., "YOLOrs: Object Detection in Multimodal Remote Sensing Imagery," in *IEEE Journal of Selected Topics in Applied Earth Observations and Remote Sensing*, vol. 14, pp. 1497–1508, 2021, doi: 10.1109/JSTARS.2020.3041316.

16. C. Dewi, R. -C. Chen, Y. -T. Liu, X. Jiang and K. D. Hartomo, "Yolo V4 for Advanced Traffic Sign Recognition With Synthetic Training Data Generated by Various GAN," in *IEEE Access*, vol. 9, pp. 97228–97242, 2021, doi: 10.1109/ACCESS.2021.3094201.

17. H. Zhou, Y. Wang and M. Ye, "A Method of CNN Traffic Classification Based on Sppnet," *2018 14th International Conference on Computational Intelligence and Security (CIS)*, 2018, pp. 390–394, doi: 10.1109/CIS2018.2018.00093.

18. V. Murugan, V. R. Vijaykumar and A. Nidhila, "Vehicle Logo Recognition using RCNN for Intelligent Transportation Systems," *2019 International Conference on Wireless Communications Signal Processing and Networking (WiSPNET)*, 2019, pp. 107–111, doi: 10.1109/WiSPNET45539.2019.9032733.

19. X. Chen, Q. Zhang, J. Han, X. Han, Y. Liu and Y. Fang, "Object Detection of Optical Remote Sensing Image Based on Improved Faster RCNN," *2019 IEEE 5th International Conference on Computer and Communications (ICCC)*, 2019, pp. 1787–1791, doi: 10.1109/ICCC47050.2019.9064409.

20. K. S. Htet and M. M. Sein, "Event Analysis for Vehicle Classification using Fast RCNN," *2020 IEEE 9th Global Conference on Consumer Electronics (GCCE)*, 2020, pp. 403–404, doi: 10.1109/GCCE50665.2020.9291978.

21. C. Jun, Y. Suhua and J. Shaofeng, "Automatic Classification and Recognition of Complex Documents Based on Faster RCNN," *2019 14th IEEE International Conference on Electronic Measurement & Instruments (ICEMI)*, 2019, pp. 573–577, doi: 10.1109/ICEMI46757.2019.9101847.

22. J. Shi, Y. Zhou and W. X. Q. Zhang, "Target Detection Based on Improved Mask RCNN in Service Robot," *2019 Chinese Control Conference (CCC)*, 2019, pp. 8519–8524, doi: 10.23919/ChiCC.2019.8866278.

23. Z. Yin, X. Wang and L. Li, "Optimization of Human Body Attitude Detection Based on Mask RCNN," *2020 8th International Conference on Orange Technology (ICOT)*, 2020, pp. 1–4, doi: 10.1109/ICOT51877.2020.9468723.

24. Z. Jie, W. Muqing and X. Weiyao, "A Two-Pathway Convolutional Neural Network with Temporal Pyramid Network for Action Recognition," *2020 IEEE 6th International Conference on Computer and Communications (ICCC)*, 2020, pp. 2448–2452, doi: 10.1109/ICCC51575.2020.9345152.

25. S. Feng et al., "CPFNet: Context Pyramid Fusion Network for Medical Image Segmentation," in *IEEE Transactions on Medical Imaging*, vol. 39, no. 10, pp. 3008–3018, 2020, doi: 10.1109/TMI.2020.2983721.

26. C. Yang, X. Lan, H. Zhang and N. Zheng, "Autonomous Tool Construction with Gated Graph Neural Network," *2020 IEEE International Conference on Robotics and Automation (ICRA)*, 2020, pp. 9708–9714, doi: 10.1109/ICRA40945.2020.9197285.

27. J. Wang and X. Hu, "Convolutional Neural Networks with Gated Recurrent Connections," in *IEEE Transactions on Pattern Analysis and Machine Intelligence*, doi: 10.1109/TPAMI.2021.3054614.

28. M. Agrawal, A. Sah, S. Shah and S. Sahana, "Fingerprint Authenticated Secure AndroidNotes" in *2021 International Journal of Technical Research & Science (IJTRS)*. ISSN Number: 2454-2024.

29. R. Hrithik Saga, T. Ashraf, A. Bingi, A. Pola, and S. Sahana, "Malignant Skin Cancer Detection Using Convolutional Neural Networking," *International Journal of Technical Research & Science*, ISSN No.: 2454-2024.

10 AI and IoT-Enabled Technologies and Applications for Smart City

Shailesh Kumar Gupta and Neha Singh
Raj Kumar Goel Institute of Technology

CONTENTS

DOI: 10.1201/9781003248750-10

10.1 INTRODUCTION

Cities will become more urbanized in the next years, and megacities with populations of more than 10 million people will emerge. These densely populated megacities will face problems in creating sustainable and cost-effective settings, improving inhabitants' quality of life, and handling nonstatic ideas that change over time. To meet these difficulties, the most up-to-date information and communication technology (ICT), as well as its associated services, is required. ICT provides a paradigm that supports long-term economic growth and superior quality of living while also confirming prudent resource management. ICT infrastructure is useful to link smart houses into a unified smart city (SC) idea. Artificial Intelligence (AI), Clouds of Things (CoT), and the Internet of Things (IoT) are important components of this idea [1–3]. Hence, data generated from smart homes will be very significant in the creation of many SC services. Figure 10.1 represents the layout of the SC in the modern era having a smart home, smart agriculture, IoT, smart health, and many more.

We are heading in that way with sophisticated IoT technologies. As connected things become more powerful, we are seeing an architectural movement away from cloud-based IoT systems and toward edge AI and embedded AI [4]. Lower latency improved privacy and the necessity to analyze data close to the source have all influenced this. For better understanding, the block diagram representation of a SC having its various features is shown in Figure 10.2.

10.1.1 INTERNET OF THINGS

The IoT is thought to be the next significant step toward the growth of internet. According to the European Union Commission's IoT action plan for Europe, the IoT will dramatically alter the way our infrastructure work in the next 5–15 years. Everyday items are being transformed into context-aware IoTs and intelligent using a mixture of the Internet and upcoming technologies such as context awareness, wireless communications, and embedded wireless sensor networks (WSNs). As a result, based on their processing capability and power usage restrictions, these IoTs will give communication and context awareness capabilities, as well as some level of pseudo-intelligence [1]. By 2020, it is expected that 7 billion people would utilize 7 trillion wireless devices, equating to more than a thousand gadgets for each individual around the globe [4], with most of them being IoT devices.

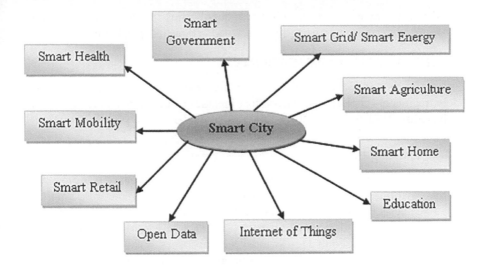

FIGURE 10.1   The layout of smart city in modern era.

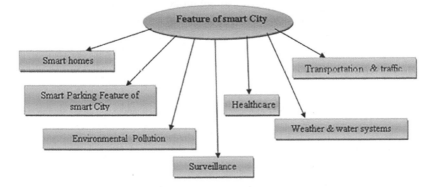

FIGURE 10.2   Features of smart city.

10.1.2 ARTIFICIAL INTELLIGENCE

Scientific research tools, toys, medical diagnostics, and robot control are just a few of AI's uses. Recommender engines, self-navigating vacuum cleaners, game engines, speech recognition, auto gearboxes, and industrial robots are just a handful of the services that use embedded AI today. Context-aware services that can handle everyday tasks like cooking eating, taking medicine, drinking, and grooming must be implemented in smart settings in the smart home sector. Hundreds of sensors must be interfaced with these systems [5]. Furthermore, they must be able to cope with huge amounts of data, which is extremely difficult for AI to learn and anticipate [6]. In general, AI-based technologies are used to offer context-aware customer services to smart homes. These systems must have the ability to learn actions from users' actions, such as at what time they walk around and do tasks in their home. When activities are learned, the system must have the ability to recognize this "learning scenario" with a high degree of certainty and perform the learned actions independently.

10.2 IOT-ENABLED APPLICATIONS

The timeline of IoT-enabled applications is depicted in Figure 10.3. The IoT is a widely fragmented application scenario that includes a broad range of uses, a few of which are listed below.

10.2.1 SMART CITIES

Smart Cities can better manage resources, become more resilient to transitory failures and calamities, and encourage efficient behavior thanks to the IoTs. Smart, as well as weather-adaptive lighting, smart parking having dynamic pricing, water/gas leakage monitoring, and automatic parking guidance, are just a few examples of how the IoTs may be used to address today's urban issues. Vision in IoT allows for physical augmentation of social sites and human activity surveillance, in addition to providing omnipresent and enhanced surveillance. This allows for a more dynamic balance between demand and supply of services [2,7].

It can help the city conserve electricity (and money) while maintaining security by preventing dark areas around people [8]. Smart tourism guarantees to provide tourists with an immediate awareness of the city, such as accessibility, crowdedness, and calmness of various locations, as well as dynamic suggestions on tours that adapt to their preferences.

10.2.2 RETAIL AND LOGISTICS

Monitoring supply chain storage conditions and inventory for tracing and payment processing based on geography or activity time in public transit, amusement parks,

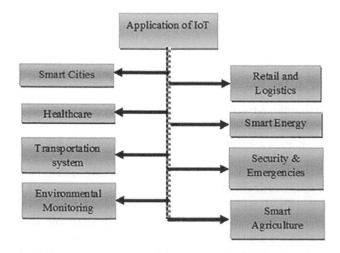

FIGURE 10.3 IoT-enabled applications.

and other applications are a few of the benefits in the use of IoT in retail/supply chain management. IoT also offers several applications in the store, such as guiding customers through the store based on a pre-determined shopping list, quick payment solutions such as automatic checking out utilizing biometrics, recognition of all potential allergens in any given item, as well as storage facilities to automate restocking operations [9].

10.2.3 HEALTHCARE

In the healthcare industry, IoT technology may be used in a variety of ways. On the one hand, they can be utilized to improve existing assisted living arrangements. Medical sensors will be worn by patients to continually measure pulse rate, oxygen level, blood pressure, the temperature of body, and blood glucose level [10,11]. Other sensors will be utilized to collect data that will be used to monitor patient behaviors in their homes. This data will be gathered locally and shared with central medical facilities, having the ability to undertake advanced monitoring remotely and respond quickly if necessary. Remote observation also expands the ability to share specialists with a larger group of individuals and patients, decreasing treatment costs [12].

10.2.4 SMART ENERGY

The smart grid (SG) is a new kind of intelligent power system that may enhance energy efficiency, minimize environmental effects, increase electricity supply safety and dependability, and minimize grid electricity transmission. Through the effective implementation of energy sensors, the incorporation of IoT technologies in the SGs may aid in defect monitoring and detection, as well as utilization monitoring [13]. Other similar ideas propose heat and energy control in houses and buildings intending to reduce energy consumption. At the power generation level, using IoT technology to gather data on energy usage may also assist manufacturing businesses to enhance their competitiveness and energy efficiency. Energy management in buildings: This application combines information systems, functional technologies, and sophisticated metering to provide real-time monitoring, reporting, and alerting to operational employees [8]. These methods are capable of providing dynamic visibility into the functioning of buildings and other facilities. Home energy management is the process of adjusting the temperature and lighting of a room based on the number of persons in the room, the time of day, the weather, and the energy cost.

10.2.5 TRANSPORTATION SYSTEM

Using sophisticated sensor, information, and network technologies, the intelligent transportation system (ITS) [14] will facilitate effective control and management of transportation. Mobile emergency command and scheduling, non-stop electronic highway tolls, vehicle guidelines violation monitoring, transportation law enforcement, anti-theft structure, lowering environmental pollution, avoiding traffic congestion, smart beaconing, reporting vehicle accidents, minimizing arrival disruptions, and so on are all possible features of intelligent transportation. All sorts of vehicles in a town (cars, buses, trains, and bicycles) are being increasingly outfitted with sensors and/or actuators, resulting in a network made up of a collection of mobile sensors. Roads and railroads, as well as carried cargo, are all outfitted with tags and sensors that provide vital data to traffic control stations. This not only enables the tracking of carried products but also for the development of novel solutions, such as allowing vehicles to better direct traffic or providing tourists with relevant transportation information. As a result, the notion of "smart cars" develops, and such data, if correctly collected and supplied, can help to make road transportation greener, smarter, and safer [9].

10.2.6 Security and Emergencies

IoT technology is used in the realm of security and emergency extensively, with a few examples such as access control of perimeter, levels of radiation, explosive dangerous gases, and liquid presence. Access control of perimeter is utilized for identifying and regulating unauthorized persons entering restricted zones, space monitoring, people and asset tracking, equipment maintenance, infrastructure, alerting, and so on. To avoid corrosion and breakdowns, the presence of liquid is utilized for liquid identification in the data centers, grounds of sensitive buildings, and warehouses [8]. The radiation levels tool is being used to monitor radiation levels in nuclear power station settings to produce leakage warnings, while the final IoT tool is used to identify gas levels as well as leakages in production environments, chemical factory surrounds, and mines. By detecting disturbances or other natural catastrophes and taking necessary steps in advance, the mixture of sensors, their autonomous management, and modeling can assist to anticipate the onset of earthquakes and volcanoes [2].

10.2.7 Environmental Monitoring

One of the most potential market sectors in the future is the use of wireless identifiable gadgets and IoT technology in nature conservation and other green activities. Bio-monitoring, soil monitoring, remote sensing, water monitoring, and pollution monitoring will all see a rise in their use in environmentally friendly initiatives throughout the world. IoT could be utilized to advance environmental programs, such as the gathering of recyclable materials for reutilization, the removal of electronic waste, and the gathering of recyclable materials for reutilization [15]. Weather monitoring includes humidity, pressure, temperature, wind speed, and rain, as well as earthquake early identification [10].

10.2.8 Smart Agriculture

A network of sensors may collect data, process it, and alert the farmer via communication infrastructure, such as SMS text communication, regarding the areas of land requiring special care. Smart seed packaging, fertilizer, and pest control devices that adapt to particular regional circumstances and identify actions are examples of this. By having information about land conditions and climatic variability, intelligent farming systems will assist agronomists in having a better grasp of plant development models and effective farming techniques. Eliminating inefficient farming circumstances will greatly boost agricultural production [16].

10.3 INTERNET OF THINGS AND ARTIFICIAL INTELLIGENCE: A LITERATURE SURVEY

A lot of developments have been done in the last few decades in smart cities having AI and IoT and continued. Many authors focused on various technologies and implemented them in various applications. Therefore, proceeding further it is necessary to go through the previous findings. This section focuses on the survey on AI and IoT-enabled technologies and applications for the SC.

10.3.1 LITERATURE SURVEY ON IoT FOR SMART CITY

The authors in [17] examined and presented the evolution, architectures, technologies and applications, and problems of smart cities. It aids the reader in comprehending the coherence of IoT-based SC development. In the development of urban infrastructure, the IoT plays a very significant role in increasing system productivity and dependability. Smart cities are more practical and sustainable as a product of the integration of big data, IoT, and cloud computing. The author discussed the many features of smart cities using IoT technology. The evolution of smart cities, their designs, applications, technology, standards, and difficulties are all extensively covered.

The authors in [18,19] performed thorough research of the relevance, use levels, and advances of IoT-based smart cities. A few difficult challenges, as well as the weaknesses smart cities based on IoT, are also recognized. The primary objective of this study is to compile prior experience from across the world to spur more efficient use of IoT in the smart cities. A lot of advanced technologies utilized in smart cities based on IoT, as well as effective incentives, were mentioned. Similarly, future problems and limits associated with the use of smart cities based on IoT are described.

The authors in [20] outline present and future SC and IoT trends and have gone through some of the IoT's flaws and how they may be solved when utilized in smart cities. The limitations and dangers related to IoT implementation and acceptance in the SC context were also recognized. The authors examined the many solutions and proposals to solve some of the difficulties of IoT and smart cities and also highlighted particularly the security challenges and issues, as part of the future work.

10.3.2 LITERATURE SURVEY ON AI FOR SMART CITY

Whether the AI is supervised or unsupervised determines how this works [10]. Target and dataset values are produced in supervised learning to train AI networks to identify particular solutions in the acquired raw data.

AI applications can help smart cities enhance, develop energy infrastructures, urban services, empower and resilient populations, among other things [21–23]. Local governments, people, and other SC stakeholders, on the other hand, face several obstacles when it comes to implementing such apps. Smart city technologies, to work, need the handling of massive amounts of data, often known as "Big Data." The three "Vs" of big data have been discussed as "high-volume, high-velocity, and/or high-variety information assets" [8], implying huge datasets processed fast through algorithms and the utilization of several data sources, including merging datasets.

The author [24] begins by defining the goals of smart monitoring, which serves as a foundation for categorizing AI, machine learning (ML) smart monitoring algorithms that may be changed to produce explainable AI in the civil engineering.

The authors in [25] present a detailed review of the literature work that shows that research work on AI for sustainability is hampered by issues such as overreliance on the historical data in machine-learning–based models, the uncertain human behavioral responses to AI-based interventions, expanding cyber security risks, negative effects of AI applications, and challenges in measuring intervention. Future AI for sustainability research should include many points as multilevel perspectives, design thinking, systems dynamics approaches, sociological and psychological considerations, and economic value factors, according to the review.

The author in [26] investigates such techniques to discover plans for the explicit implementation of AI and robots. An internet keyword search yielded a total of 12 case studies covering cities of varying sizes throughout the world. The terms "artificial intelligence" and "robot," which represent robotics and related terminology, were used in the search. According to the data, the Global North presently has the most contemplated deployment of AI and robots in SC's development, while nations in Global South are becoming more represented. Numerous cities in Canada and Australia are dynamically working to design and develop AI and robots, with Moscow having one of the most detailed plans.

The authors in [27] investigate the role of AI, ML, and deep reinforcement learning in the growth of the smart cities. These approaches described above are effective in determining the best policy for a variety of smart-city–related challenges. It presents detailed in-depth information of the applications of previous practices in cyber security, ITSs, efficient SGs, effective usage of unmanned aerial vehicles to ensure smart healthcare system and the best 5G and beyond 5G (B5G) technologies in communication services in a SC in this survey.

10.4   AI AND IOT-ENABLED TECHNOLOGIES

There are various technologies [4,9,17,20,28,29,30] that are used to define IoT, but the range-wise technologies are shown in Figure 10.4. Range-wise IoT technologies are classified into three categories: short, middle, and wide range, which are further classified as shown in Figure 10.4.

10.4.1   Radio Frequency Identification (RFID)

Radio frequencies are used by RFID to send and receive data. In RFID communication, two kinds of devices are used: the reader and the tag [29]. When a tag approaches the reader, the exchange of knowledge happens following authorization; while tag absorbs reader's energy, passive tags are those that rely on the reader for their power; active tags, on the other hand, do not. RFID may function on multiple

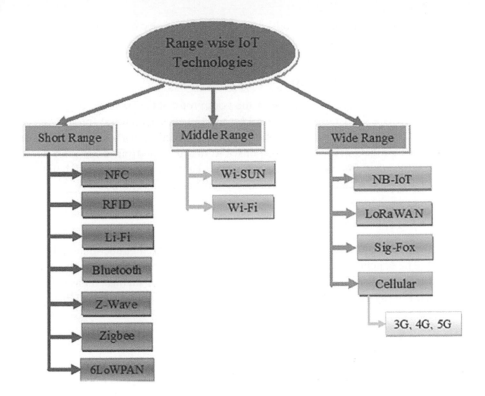

FIGURE 10.4   Range-wise classification of IoT technologies.

frequencies within the radio frequency band in the middle of 125 kHz–928 MHz, depending on the standard, and can be utilized across short distances.

10.4.2   NEAR-FIELD COMMUNICATION (NFC)

NFC is comparable to the RFID, except it lacks tags and scanners that RFID does. Unlike RFID, both devices that wish to interact via NFC must be powered, and unlike RFID, data transmission and reception can occur in both ways. Unlike RFID, which may not be utilized for control activities, NFC may be used to control and set up devices. NFC uses frequencies that are comparable to RFID; however, it is only utilized for extremely short distances.

10.4.3   BLUETOOTH

Bluetooth is a low-energy standard that can handle an infinite number of nodes, making it useful in IoT applications [30]. The set of rules is intended for short-range, small-bandwidth communication in a network configuration that allows devices to quickly depart and enter. Because it contains a master device at the heart of the communications platform, Bluetooth offers the star topology out of the box. It uses the 2.4-GHz ISM band and may transmit data at up to 2 Mbps.

10.4.4 ZENSYS WAVE (Z-WAVE)

Z-Wave, also known as Zensys Wave, is the poor protocol designed for use in home automation. It is a short-range, less-speed set of rules that operates at frequencies of 868 and 900 MHz. It works in a master-slave model, meaning that a master may have various slave gadgets that can give responses to instructions from the master node. As a result, it's ideal for applications that require a central control part and data from several sensing units, like smart healthcare systems and smart homes.

10.4.5 LIGHT FIDELITY (LI-FI)

To communicate data, Li-Fi (Light Fidelity) employs visible light rather than radio frequencies. The benefit of adopting Light Fidelity over RF communication is that it may take advantage of existing lighting systems, resulting in power savings. It has been utilized in parking systems and provides very high data transfer rates over short distances.

10.4.6 WIRELESS FIDELITY (WI-FI)

Wi-Fi is a wireless technology that uses the 2.4 and 5 GHz bands to deliver great-speed internet access over a short area. Many SC applications employ Wi-Fi because it provides a fully prepared interface for smartphones, laptops, and other wearable devices.

10.4.7 ZIGBEE

The ZigBee standard was created as a low-cost and power WSN standard and has now evolved to be utilized in the IoT. In a multi-hop data transfer system, the ZigBee communications protocol runs in the 868/915 MHz/2.4 GHz spectrum and delivers modest data transfer speeds having distances equivalent to Wi-Fi. Because ZigBee radios are inexpensive, they are a popular protocol among smart healthcare device makers and smart homes.

10.4.8 WIRELESS SMART UTILITY NETWORK (WI-SUN)

The Institute of Electrical and Electronics Engineers (IEEE)-approved Wi-SUN is utilized in the field area network for utility monitoring, distribution automation in applications, including gas, electricity, and water, as well as demand response systems for utility-based activities.

10.4.9 CELLULAR TECHNOLOGIES

3G, 4G, and 5G communications are all referred to as cellular technology. They are the most widely used IoT-enabling techniques, alongside Bluetooth and ZigBee. In comparison to other standards, cellular communication delivers a large data rate and allows more content-rich operations. Cellular bands with very faster speeds range from 600 MHz to 80 GHz, depending on the method.

10.4.10   LONG RANGE WIDE AREA NETWORK (LoRaWAN)

LoRaWAN is low-power wide area network (LPWAN) which consists of a back-end network server and several gateways and end devices. Connection to the cloud is provided via the back-end server. When it comes to transferring data to cloud, end devices don't having a set relationship with single gateway &may transmit data to the numerous gateways.

10.4.11   LOW-POWER WIRELESS PERSONAL AREA NETWORKS (6LoWPAN)

The Internet Engineering Task Force (IETF) established 6LoWPAN, which stands for IPv6 over less-power networks, particularly for IoT applications to enable internet access to tiny devices. It's a network based on IP that uses IPv6 for communication. It is a small-ranged network that uses ISM bands to communicate.

10.4.12   SIGFOX TECHNOLOGY

SigFox is a proprietary standard created by SigFox Inc., a company based in France. It performs ultra-narrowband bidirectional communication at low rates using unlicensed bands. SigFox, like LoRaWAN and 6LoWPAN, has a similar design and is a common LPWAN in the IoT area, enabling a sufficiently long communication range of up to 50 km. Smart lighting, building security, and environmental monitoring are all uses for SigFox.

10.4.13   NARROW BAND IoT (NB-IoT)

NB-IoT is a kind of LPWAN that uses the global system for long-term evolution (LTE) and mobile communications (GSM) bands to communicate. Because it is a bare-bones version of LTE, it may run on the same hardware having an upgrade in software. It can link up to 100,000 devices per cell.

A comparison study among IoT technologies is given in Table 10.1.

10.5   THINGS TO BE REMEMBERED IN AI AND IOT-ENABLED SMART CITY

This section demonstrates the debate by providing analysis. In the application of IoT in smart cities, we perform flaws, strengths, prospects, and threat analysis which examine the benefits that IoT provides for the SC, the drawbacks that exist in recent execution scenario, the threats that IoT usage to smart cities faces, and the prospects for future work in this zone.

10.5.1   STRENGTHS

The advantages of IoT smart cities include that they increase the quality of life for city residents while also lowering operating costs and enabling towns to be more

TABLE 10.1

Comparison among IoT Technologies

	Standard	Frequency	Maximum Data Range	Maximum Range	Mesh Network Support	Overall Cost
NFC	–	13.56 MHz	106 or 212 or 424 kbps	<10 cm	No	Low
Bluetooth	802.15.1	2.4 GHz	2 Mbps	200 m	Yes	Low
Z-Wave	802.15.4	908.4 MHz	100 kbps	100 m	Yes	Moderate
ZigBee	802.15.4	800 and 900 MHz, 2.4 GHz	250 kbps	100 m	Yes	Moderate
6LoWPAN	802.15.4	868 and 915 MHz, 2.4 GHz	250 kbps	10 m	Yes	Decreasing
Wi-Fi	802.11a, b, g, n, ac	2.4 and 5 GHz	Up to 1 Gbps	40 m	Yes	High
NB-IoT	3GPP Release 13/14	Various	250 kbps	20 km +	No	Moderate
LoRaWAN	802.15g	470–510 MHz 865–925 MHz	27 kbps	10 km +	No	Low
SigFox	Binary phase-shift keying	100 Hz–900 MHz	<100 bps	100 m +	Yes	Low
Cellular	3GPP	200 KHz– 900 MHz	50–200 Mbps (LTE)	35 km (GSM), 200 km (HSPA)	No	High

sustainable. Sensors and gadgets can be placed in a city to provide an outline of condition of city's key services, like electric, water, transportation, gas distribution, and monitoring of crime, to mention a few. Furthermore, because of the dispersed nature of IoT structures and flexible designs that permits fluidity using the mobility of sensor units, upgrading and expanding existing deployed systems is quite in expensive. Furthermore, this distributed design makes such systems extremely fault tolerant, boosting deployment dependability and providing self-healing in uses like power systems.

10.5.2 WEAKNESSES

IoT in smart cities has several technological problems. As noted in the chapter, the current deployment situation comprises a variety of different types of technologies concerning networks, software frameworks, and hardware platforms, that does not necessarily operate well together. The IETF, the IEEE, the European Telecommunications Standards Institute, and other organizations have contributed communication standards, identification, discovery of networks, and management of devices, among other things. Nevertheless, the sheer number of "standards," many of which are incompatible with one another, hasn't completely solved the interoperability problem, which might stymie IoT system expansion without a substantial redesign of system components.

10.5.3 OPPORTUNITIES

In terms of reducing vulnerabilities and providing new municipal services, IoT in smart cities offers numerous potentials to the researchers and companies alike. The data collected by sensors in IoT devices can give a comprehensive perspective of city's status, enabling the development of new applications and services using big data techniques. This heterogeneous data presents a fantastic opportunity for data analytics researchers to build innovative data science algorithms for maintenance delivery. To make IoT deployment faster and cheaper, there is a high monetary value placed on the progress and use of computationally inexpensive encryption methods, efficient data storage systems, and networking techniques. Another potential for the IoT researchers working on SCs is the progress of new sensor technologies. Creation of improved, greater efficiency, and reduced cost sensors would help construction of IoT infrastructure, enabling broad usage.

10.5.4 THREATS

There are numerous dangers associated with IoT devices used in smart cities, including user trust difficulties, concerns of privacy on account of network assaults, potential data theft, and so on. The most important interests of IoT applications are privacy and security. With such a personalized interaction mechanism between devices and people as in smart cities, data theft, the risks of leakages and privacy breaches are significant, and this is a constant interest for both service users and providers. Several cyber-attacks against SC's infrastructure have highlighted the skill's susceptibility to cyber-attacks, as well as the repercussions for the general public. Because IoT devices generally lack adequate computational capabilities, conventional security processes and techniques like routing, authentication of access, and networking may not be sufficient or viable in various IoT implementations. This has increased security and privacy issues for stakeholders in IoT ecosystem. Customers' lack of confidence in SC apps may be exacerbated as a result of this.

10.6 CHALLENGES FOR DEPLOYMENT OF IOT IN SMART CITY

The deployment of IoT is dependent on several important variables. Understanding the implementation issues may aid management in the design, direction, and control of SC services [28,31,32]. To deliver higher quality services, policymakers might prepare for IoT deployment. The problems for this study were chosen after examining the relevant literature (given in Table 10.2).

The interdependence is between semantic features and the overlap of urban regions. There are two types of big data research problems for smart cities: business

TABLE 10.2

Challenges in the Deployment of IoT in SCs

Challenges	Description	Resource
IoT device compatibility and connectivity are lacking	The difficulty of various manufacturers' IoT devices to interact and exchange data.	[36,37]
IoT gadgets with poor design	Poorly planned and deployed smart city systems may have a detrimental impact on network resource usage and overall smart city operations.	[35,38]
The capacity to cooperate (homogenous networks)	The goal of interoperability is to have all the IoT devices work together in an integrated software system.	[39,40]
Standards are lacking	Improper regulatory requirements make it difficult to organize and manage large amounts of unstructured data.	[31,41]
Inability to use the internet (developers and designers)	For IoT to work, you need to know how to utilize the internet. There is a scarcity of people to design, implement, & manage IoT systems in the market, according to IoT firms.	[32,42]
Concerns about data security and privacy	IoT devices handle data that contains private and confidential information about people's behavior. Stakeholders can be profiled as a result of a poor security procedure or a data leak.	[43–45]
Implementation of a difficult networking strategy	Several components connected to the network places a major demand on it, &implementation of network is the key issue in this field.	[46,47]
Insufficient analysis and updating of data collection	In terms of sensors and security, the requirement for updated software and hardware of an IoT system plays key role in enforcing data-specific regulations and detecting abnormalities (anomalous data) and traffic patterns.	[48]
Viability in terms of money	Because the IoT's application uses several actuating and sensing components and devices, the payback period and cost will be a major consideration.	[28,49,52]
Infrastructure mobility in cities	For municipal infrastructure to be mobile in the SCs, IoT infrastructures that can cope with the mobile data sources are required.	[13,50,51,53]

problems (planning, cost, sustainability, and cloud computing integration) and technological challenges [33,34] (privacy, data formats, data analytics, and Quality of Service (QoS)).

Design and Implementation Costs—To move into the cloud and big data paradigm, smart IoT necessitates the purchase and installation of new equipment and applications. The majority of the city's administrative processes will need to be altered as a result. It is still a worry to place the sensors in the proper areas without compromising people's convenience or privacy.

Heterogeneity—Each IoT system has its own set of devices, technology, software, and platform. All of this information should be combined and analyzed on the cloud. IoT integration and device connectivity are hampered by interoperability concerns. Although standards exist, they have not yet fully matured to handle the various IoT systems.

Cyber security—In cyber-physical systems, an intrusion detection system is critical. Because the internet is the IoT's backbone, cyber assaults are widespread, and hackers attempt to steal citizen data. Hence, it is needed to design, develop, and build IoT infrastructure that can withstand such attacks. According to recent research of crime data in Japan, urbanization is one of the major causes of crime in the city.

Dynamic adoption—IoT technologies are maturing daily, making it difficult for integration of new IoT systems into current IoT systems with little modifications. For the long-term viability of SCs, IoT sustainability is necessary by finding the system that acknowledges the noble technical viability.

Security concerns—Because it includes a large amount of citizen data to be managed and made available on the network, smart IoT should be secure. The IoT network's security and privacy [31] is a key problem in the SC project. Even though there were several technical options accessible, infiltration is unavoidable. The IoT architecture must include both proactive and reactive measures. Ijaz et al. divided SC security concerns into three categories [8,35].

a. **IoT technologies:** In use of RFID (tag's abuse, denial of service, jamming, and spoofing), WSN (tag disabling or killing, confidentiality integrity issues, battery exhaustion), smartphones (Bluetooth/GPS/Wi-Fi threats, malware, Botnets), M2M communications (DoS, attacks on side channel, protocol attacks, MITM attacks),

b. **Governmental factors:** Utility (misuse and exploitation of data), essential organization (issues in the health area, telecommunications, electricity, and power supply), and smart mobility (individual privacy and location),

c. **Socioeconomic factors:** In the smart communication (cyber security and data integrity), banking (online phishing, cybercrime, online frauds), personal privacy (social media, smartphone usage, location privacy), and e-commerce concerns.

10.7   CONCLUSION

This chapter covers a broad range of AI and IoT-enabled SC, technologies used, and applications. The IoT is the most effective approach to prepare a city smart. Indeed, IoT may be used in a variety of scenarios, including building monitoring using passive WSNs, environmental monitoring, smart parking, waste management, decreasing CO_2 emissions, and autonomous driving. Such objectives need a massive number of interconnected items. Indeed, the number of linked items is rapidly increasing, with over 60 billion associated objects expected to be implemented in smart cities by 2022. However, such a large quantity poses various dangers and privacy concerns. We offered an outline of IoT related to smart cities and addressed how it might improve the smartness of a city in this chapter. The final section will discuss the challenges of adopting AI and IoT technologies for smart cities, as well as the conclusion and future difficulties. We want to examine the many solutions and proposals to solve some of the difficulties of IoT and smart cities highlighted in this chapter, particularly the security challenges and issues, as part of our future work. This chapter is intended to help academics working in this subject by giving a detailed introduction to the usage of AI and IoT in smart cities.

REFERENCES

[1] J. Liu and W. Tong, Dynamic services model based on context resources in the internet of things, *6th International Conference on Wireless Communications Networking and Mobile Computing (WiCOM), 2010, 23–25 September.* 2010. doi: 10.1109/WICOM.2010.5601423.

[2] M. Alam, M. Reaz and M. Ali, A Review of Smart Homes—Past, Present, and Future, *IEEE Transactions on Systems, Man, and Cybernetics, Part C (Applications and Reviews),* 42, 6, pp. 1190–1203, November 2012. doi: 10.1109/TSMCC.2012.2189204.

[3] A. Somov, C. Dupont, and R. Giaffreda, Supporting Smart-City Mobility with Cognitive Internet of Things, in *Future Network and Mobile Summit (FutureNetworkSummit), 2013.* IEEE, 2013, pp. 1–10.

[4] T. Rahim, Technologies for the Wireless Future, *Wireless World Research Forum (WWRF) (Wiley-WWRF Series Book 8),* 1st Edition, Kindle Edition, vol. 2, 2008.

[5] D. J. Cook, Learning Setting-Generalized Activity Models for Smart Spaces, *IEEE Intelligent Systems,* 99, 1, 32–38, 2011. doi: 10.1109/MIS.2010.112.

[6] M. Dominici, G. Zecca, F. Weis and M. Banatre, Physical Approach in Smart Homes A Proposition and a Prototype, *3rd Conference on Smart Space and Community (RuSMART'10), St Petersburg, Russia.* 2010.

[7] N. Komninos, P. Tsarchopoulos and C. Kakderi, New Services Design for Smart Cities: A Planning Roadmap for User-Driven Innovation *Proceedings of the 2014 ACM International Workshop on Wireless and Mobile Technologies for Smart Cities, Philadelphia, PA, USA,* pp. 29–38, 11–14 August 2014. doi: 10.1145/2633661.2633664.

[8] W. M. Kang, S. Y. Moon and J. H. Park, An enhanced security framework for home appliances in smart home, *Human-centric Computing and Information Sciences,* 7, 1, Article 6, 2017.

[9] S. Vongsingthong and S. Smanchat, Internet of things: A review of applications and technologies. *Suranaree Journal of Science and Technology,* 21, 4, 359–374, 2014. doi: 10.14456/sjst.2014.38.

[10] C. C. Andréda, P. F. Cristian, E. Björn, D. B. D. Silva, R. R. Rosa, Internet of health things: Toward intelligent vital signs monitoring in hospital wards, *Artificial Intelligence in Medicine,* 89, pp. 61–69, July 2018. doi: 10.1016/j.artmed.2018.05.005.

[11] T. Mashiko, Big Data, IoT, AI, and Smart Cities, *Telecommunications Policies of Japan, Advances in Information and Communication Research,* vol 1, Springer, ISBN 978-981-15-1032-8 ISBN 978-981-15-1033-5 (eBook) pp. 181–200, January 2020. doi: 10.1007/978-981-15-1033-5.

[12] N. Janmenjoy, V. Kanithi, D. Paidi, N. Bighnaraj, M. Subhashree, S. Tripti, and M. Manohar, Intelligent Computing in IoT-Enabled Smart Cities: A Systematic Review *Green Technology for Smart City and Society, Lecture Notes in Networks and Systems,* vol. 151, December 2020. doi: 10.1007/978-981-15-8218-9_1.

[13] M. Mital, V. Chang, P. Choudhary, A. Papa and A. Pani, Adoption of Internet of Things in India: a test of competing models using a structured equation modeling approach *Technological Forecasting and Social Change,* vol. 136, pp. 339–346, 2018. doi: 10.1016/j.techfore.2017.03.001.

[14] X-F Xieand, Z-J Wang, Integrated in-vehicle decision support system for driving at signalized intersections: A prototype of smart IoT in transportation, *Transportation Research Board 96th Annual Meeting, Washington DC, United States*, 2017.

[15] H. Arasteh, V. Hosseinnezhad, V. Loia, A. Tommasetti, O. Troisi, M. Shafie-khah and P. Siano, IoT-based smart cities: A survey *IEEE 16th International Conference on Environment and Electrical Engineering (EEEIC)*. IEEE, New York, 2016. doi: 10.1109/EEEIC.2016.7555867

[16] M. Miraz, M. Ali, P. Excell, and R. Picking, Internet of Nano-Things, Things and Everything: Future Growth Trends, *Future Internet*, 10, 8, 68, 2018, doi: 10.3390/fi10080068.

[17] K. Saravanan, E. G. Julie and Y. H. Robinson, Smart Cities & IoT: Evolution of Applications, Architectures & Technologies, Present Scenarios & Future Dream In: Balas V., Solanki V., Kumar R., Khari M. (eds.), *Internet of Things and Big Data Analytics for Smart Generation. Intelligent Systems Reference Library*, vol. 154. Springer, Cham. doi: 10.1007/978-3-030-04203-5_7.

[18] A. A. Jaafar, H. K. Sharif, I. M. Ghareb and D. N. A. Jawawi, Internet of Thing and Smart City: State of the Art and Future Trends, *Advances in Computer Communication and Computational Sciences, Advances in Intelligent Systems and Computing Series (AISC)*, vol. 760, 19 August 2018, doi: 10.1007/978-981-13-0344-9_1.

[19] N. Janmenjoy, V. Kanithi, D. Paidi, N. Bighnaraj, M. Subhashree, S. Tripti, and M. Manohar, Intelligent Computing in IoT-Enabled Smart Cities: A Systematic Review, *Green Technology for Smart City and Society,* Lecture Notes in Networks and Systems vol. 151, 2021, doi: 10.1007/978-981-15-8218-9_1.

[20] V. Albino, U. Berardi and R. M. Dangelico, Smart Cities: Definitions Dimensions Performance and Initiatives, *Journal of Urban Technology*, vol. 22, pp. 3–21, January 2015.

[21] A. Pacheco, P. Cano, E. Flores, E. Trujillo and P. Marquez, A Smart Classroom Based on Deep Learning and Osmotic IoT Computing In *Proceedings of the 2018 CongresoInternacional de Innovación y TendenciasenIngeniería (CONIITI)*, Bogota, Colombia, 3–5 October 2018. doi: 10.1109/CONIITI.2018.8587095.

[22] D. Devin, V. V. Anne Fleur, T. Tjerk, T. Paola and K. Maria, Artificial Intelligence in Smart Cities and Urban Mobility, *IPOL | Policy Department for Economic, Scientific and Quality of Life Policies, PE 662.937*, July 2021.

[23] A. Nikitas, K. Michalakopoulou, E. T. Njoya and D. Karampatzakis, Artificial intelligence, transport and the smart city: Definitions and dimensions of a new mobility era, *Sustainability*, vol. 12, issue 7, 1 April 2020.

[24] D. Luckey, H. Fritz, D.Legatiuk, K. Dragos and K. Smarsly, Artificial Intelligence Techniques for Smart City Applications. In: Toledo Santos E., Scheer S. (eds.), *Proceedings of the 18th International Conference on Computing in Civil and Building Engineering*. ICCCBE 2020, Lecture Notes in Civil Engineering, vol. 98, Springer, Cham, 2021. doi: 10.1007/978-3-030-51295-8_1.

[25] R. Nishant, M. Kennedy, & J. Corbett, Artificial Intelligence for Sustainability: Challenges, Opportunities, and a Research Agenda, *International Journal of Information Management*, vol. 53, August 2020.

[26] G. Oleg and T. Mary, Artificial Intelligence and Robotics in Smart City Strategies and Planned Smart Development, *Smart Cities 2020*, 3, 4, pp. 1133–1144, 2020. doi: 10.3390/smartcities3040056.

[27] U. Zaib, T. FadiAl, M. Leonardo, and G. Roberto, Applications of Artificial Intelligence and Machine learning in smart cities, *Computer Communications*, vol. 154, pp. 313–323, 15 March 2020. doi: 10.1016/j.comcom.2020.02.069.

[28] S. Albishi, B. Soh, A. Ullah and F. Algarni, Challenges and Solutions for Applications and Technologies in the Internet of Things, *Procedia Computer Science*, vol. 124, pp. 608–614, 2017. doi: 10. 1016/j.procs.2017.12.196.

[29] H. Badis, K. Rida, Z. Sherali, F. Achraf and K. Lyes, Internet of Things (IoT) Technologies for Smart Cities *IET Networks*, 7, 1, September 2017. doi: 10.1049/iet-net.2017.0163.

[30] S. S. Abbas, S. S. Daniel, K. Anup and E. Adel, IoT in Smart Cities: A Survey of Technologies, Practices and Challenges, *Smart Cities 2021*, 4, pp. 429–475. doi: 10.3390/smartcities4020024.

[31] R. Roman, J. Zhou and J. Lopez, On the features and challenges of security and privacy in distributed Internet of Things, *Computer Networks*, vol. 57, issue 10, pp. 2266–2279, 5 July 2013. doi: 10.1016/j.comnet.2012.12.018

[32] B. R. Stojkoska and K. Trivodaliev A review of Internet of Things for smart home: challenges and solutions, *J Clean Prod*, vol. 140, issue 3, pp.1454–1464, 1 January 2017. doi: 10.1016/j.jclepro. 2016.10.006, 2017.

[33] K. I. Haleem, K. M. Imran and K. Shahbaz, Challenges of IoT Implementation in Smart City Development*Lecture Notes in Civil Engineering,* vol. 58, Smart Cities— Opportunities and Challenges Select Proceedings of ICSC 2019, pp 475–486., ISBN 978–981-15–2544–5 ISBN 978–981-15–2545-2 (eBook) doi: 10.1007/978–981-15–2545–2.

[34] S. KnudErik, P.Lynggaard, I. Windekilde and A. Henten, How IoT, AAI can contribute to smart home and smart cities services: The role of innovation, *25th European Regional Conference of the International Telecommunications Society (ITS): Disruptive Innovation in the ICT Industries: Challenges for European Policy and Business,* Brussels, Belgium, 22nd-25th June 2014.

[35] C. Tankard, The security issues of the Internet of Things, *Computer Fraud & Security*, vol. 2015 issue 9, pp. 11–14, 2015. doi: 10.1016/s1361-3723 (15)30084–1.

[36] P. Asghari, A. Rahmani and H. Javadi, Internet of Things applications: a systematic review, *Computer Networks*, vol. 148, pp. 241–261, January 2019. doi: 10.1016/j. comnet.2018.12.008.

[37] L. Atzori, A. Iera and G. Morabito, The Internet of Things: a survey, *Computer Networks*, vol. 54, issue 15, pp. 2787–2805, October 2010. doi: 10.1016/j.comnet.2010.05.010.

[38] J. Jin, J. Gubbi, S. Marusic and M. Palaniswami, An information framework for creating a smart city through Internet of Things, *IEEE Internet of Things Journal*, vol. 1, issue 2, pp. 112–121, April 2014.DOI: 10.1109/JIOT.2013.2296516.

[39] O. Bello and S. Zeadally, Toward efficient smartification of the Internet of Things (IoT) services, *Future Generation Computer Systems*, vol. 92, pp. 663–673, March 2019. doi: 10.1016/j.future.2017.09.083.

[40] I. Yaqoob, I. Hashem, A. Ahmed, S. Kazmi and C. Hong, Internet of Things forensics: recent advances, taxonomy, requirements, and open challenges, *Future Generation Computer Systems*, 92, 265–275, March 2019. doi: 10.1016/j.future.2018.09.058.

[41] M. Rathore, A. Paul, A. Ahmad, and G. Jeon IoT-based big data, *International Journal on Semantic Web and Information Systems (IJSWIS)*, 13, 1, 28–47, 2017. doi: 10.4018/ijswis.2017010103.

[42] P. Boer, A. van Deursen and T. van Rompay, Accepting the internet-of-things in our homes: the role of user skills, *Telematics and Informatics*, 36, 147–156, March 2019. doi: 10.1016/j.tele.2018.12.004.

[43] G. Sagirlar, B. Carminati and E. Ferrari, Decentralizing privacy enforcement for Internet of Things smart objects, *Computer Networks*, 143, 112–125, October 2018. doi: 10.1016/j.comnet.2018.07.019.

[44] S. Sicari, A. Rizzardi, L. Grieco and A. Coen-Porisini, Security, privacy and trust in Internet of Things: The road ahead, *Computer Networks*, 76, 146–164, January 2015. doi: 10.1016/j.comnet.2014.11.008.

[45] R. Weber, Internet of Things: privacy issues revisited, *Computer Law Security Review*, vol. 31, issue 5, pp. 618–627, October 2015. doi: 10.1016/j.clsr.2015.07.002.

[46] W. Serrano, Digital systems in smart city and infrastructure: Digital as a service, *SmartCities*, 1, 1, 134–153, November 2018. doi: 10.3390/smartcities1010008.

[47] E. Ahmed, I. Yaqoob, I. Hashem, I. Khan, A. Ahmed, M. Imranand and A. Vasilakos The role of big data analytics in Internet of Things, *Computer Network*, 129, 459–471, December 2017. doi: 10.1016/j.comnet.2017.06.013.

[48] N. Ismail, The impact of the Internet of Things (IoT), *Information Age*, 24 July 2017. https://www.information-age.com/impact-internet-things-iot-123467503.

[49] L. Liu, IoT and a sustainable city, *EnergyProcedia*, 153, pp. 342–346, October 2018. doi: 10.1016/j.egypro.2018.10.080.

[50] M. Rathore, A. Paul, A. Ahmad, and G. Jeon, IoT-Based Big Data: From Smart City towards Next Generation Super City Planning, *International Journal on Semantic Web and Information Systems (IJSWIS)*, 13, 1, pp. 28–47, 2017. doi: 10.4018/ijswis.2017010103.

[51] P. Marques, D. Manfroi, E. Deitos, J. Cegoni, R. Castilhos, J. Rochol, E. Pignaton and R. Kunst, An IoT-based smart cities infrastructure architecture applied to a waste management scenario, *Ad Hoc Network*, 87, 200–208, 2019. doi: 10.1016/j.adhoc.2018.12.009.

[52] Sagar, R. H., T. Ashraf, A. Sharma, K. S. Raj Goud, S. Sahana, and A. K. Sagar. Revolution of AI-Enabled Health Care Chat-Bot System for Patient Assistance. In *Applications of Artificial Intelligence and Machine Learning*, pp. 229–249. Springer, Singapore, 2021.

[53] Anand, A., Mishra, S.P. and Sahana, S. Assistive Devices and IoT in Healthcare Functions. *Deep Learning and IoT in Healthcare Systems: Paradigms and Applications*, 2021, 103.

11 Blood Cancer Classification with Gene Expression Using Modified Convolutional Neural Network Approach

Nidhi Gupta
Raj Kumar Goel Institute of Technology

Akhilesh Latoria
AURO University

Akash Goel
Raj Kumar Goel Institute of Technology

CONTENTS

11.1 INTRODUCTION

The emerging technique in the field of microarray technology helps scholars to determine a no of genes simultaneously achieving essential information about the functions of cell [1]. The information collected during the study can be utilized in the prognosis and diagnosis of cancer. Although, to extract characteristics of gene expression, data from large dataset and gene selection process remain a difficult process due to the presence of noise. To overcome the difficulty, a technique is required to select an appropriate set of genes having high classification accuracy [2].

This technique will save computational expenses as well as permit physicians to recognize a small fraction of the genes that are biologically specific to particular cancers. In addition, an efficient method can assist in the early diagnosis and detection drugs for cancer patients [3]. The sample of blood for normal and leukemia is shown in Figure 11.1.

As shown in Figure 11.1, blood cells are the composition of plasma along with three distinct cells (white, red, and platelets). Each one is responsible for a different task. The RBC (red blood cells) carry oxygen from lungs to the tissues and vice versa [4], whereas the white blood cells are responsible to fight against disease. The platelets are responsible to control bleeding. On the right side of Figure 11.1, leukemia is a blood cancer cells, where the growth of the white blood cell increases rapidly and interferes with the operation performed by platelets and RBC [5]. There are two main categories of leukemia that appear in the blood known as lymphocytic leukemia, which is caused due to lymphoid and other one is xylogenous leukemia, which appears due to myeloid cells. The growth of these cells increases rapidly. Therefore, it is essential to treat these cells immediately. Therefore, an atomized system needs to be created or designed so that disease can be detected at the early stage [6]. The

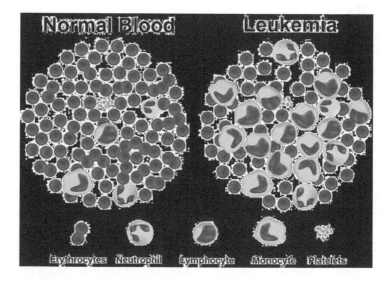

FIGURE 11.1 Normal and leukemia blood samples.

selection of genes can be performed in three distinct categories such as (i) the filter, (ii) wrapper, and (iii) embedded approaches [7].

The remaining study is divided into four sections: Section 11.2: Review of Literature related to the study in the field of blood cancer detection is presented. Section 11.3: The detailed description of proposed work and research methodology. Section 11.4: It includes discussion in terms of evaluated performance parameters. In Section 11.5: Conclusion with way forward is presented.

11.2 RELATED WORK

Recently, particle swarm optimization (PSO) algorithms are being used extensively by researchers to propose answers for the issue of gene determination. Alba in her study analyzed the utilization of PSO, SVM (support vector machine), and GA (genetic algorithm) as high-dimensional microarray information classifiers. Further developed PSO, geometric PSO is proposed to show comparative analysis with GA on six distinct public cancer data sets [8]. A gene selection approach in integration with PSO and GA as an optimization approach and SVM as a classification approach has been presented by [9]. The plan has been examined on three benchmark dataset of gene expressions that includes leukemia colon and breast cancer data.

The intention or planning of designing the model which is hybrid in nature in the proposed approach is to avoid it from falling into the local optima swarm optimization in integration with SVM classifier for the suitable determination of the close to ideal subset of data information genes which is related to cancer grouping and classification [9]. In the proposed study, Mohamad et al. [10] enhanced binary particle past and modified the particle velocity as well as positions by updating the existing rules. Gao ct al. [11] have presented an integrated approach for gene selection using SVM approach in combination with information gain (IG). Initially, the redundant genes have been filtered using IG approach. Further, to reduce noise from the dataset, SVM approach was applied. After that, the obtained output from SVM is applied to the Library for Support Vector Machines (LIB-SVM) classifier. The results obtained with IG SVM approach performed well with an accuracy of 90.32% [11]. Dwivedi has used a supervised learning approach to differentiate acute lymphoblastic and myeloid leukemia. Artificial Neural Network (ANN) has been used as a classification approach and found best among five different machine learning approaches with the accuracy level up to 98% and zero error in acute lymphoblastic leukemia (ALL) [12]. Sun et al. [13] have presented a new feature selection approach using a rough set and entropy approach for the detection of cancer from gene expression.

Further, some properties and relationships between these metrics are analyzed, which helps to understand the content information of neighborhood dynamic frameworks. At last, the Fisher scoring strategy was utilized to eliminate preliminarily unrelated genes subsequently altogether decreasing complexity, alongside the desired selection of features with lower computational complexity was proposed to upgrade the performance designed for cancer classification systems [13].

Methodology

Under this section, this chapter describes the structure and working steps for Blood Cancer Classification (BCC) model with gene expression using the Modified

Convolutional Neural Network (MCNN) approach. We will focus on introducing a modified MCNN approach in two different manners such as:

11.2.1 SPEEDED-UP-ROBUST-FEATURE–BASED MCNN

1. Firstly, we introduce the concept of extremal feature descriptor to extract the exact pattern feature form gene microarray data that may be normal or tumor tissue data. By utilizing the concept of external feature descriptor, we can pass only the exact pattern features of gene data and the chances of classification accuracy will be improved but still any problem remains, then we introduce the next concept with the utilization of PSO as a feature selection approach.

11.2.2 OPTIMIZED SURF-BASED MCNN

1. This is the second step of proposed BCC model that includes the feature selection process based on the optimal fitness function using the PSO as feature optimization approach.

Figure 11.2 depicts the outline of the proposed BCC model strategy with four different stages. The subsections highlight the role of each stage and the modifications performed at each stage.

The first part of the BCC model is the dataset of GENE expression. The collection of dataset of gene expression form Kaggle repository (availability link: Reference). This dataset is developed and collected by Golub et al. [14] for the academic research and it contains the classification of the gene expression of data for the blood cancer via DNA microarray. Basically the given dataset is used to classify patients with blood cancer in terms of acute myeloid leukemia (AML) and ALL. The overall dataset will be categorized into two parts called data testing and training of the data, where training dataset contains total 38 samples test dataset contains 34 samples for the simulation odd-designed BCC model. The blood cancer dataset contain ALL and AML samples collected from patient's bone marrow and peripheral blood. Further, the rescaling of the sample intensity values is being done in the pre-processing phase. The pre-processing is a basic and important need of the BCC model to prepare the dataset according to the requirements for the categorization of AML and ALL through the dataset of gene expression, the normal pre-processing involves one or more below mentioned processes to design a BCC model for the classification of ALL and AML cancer types:

- **Filtration:** The gene expression data filtration is one of the basic need to make noise and disturbance-free data which excludes a data from the dataset if behave like noise. Noise filtering is normally utilized to decrease the presence of noise within the gene data of ALL and AML blood cancer and creates a noise-free data with less information loss. So, filtering process of various data of gene expression is utilized for the estimate of level regarding the present noise and used filtration algorithm is composed as:

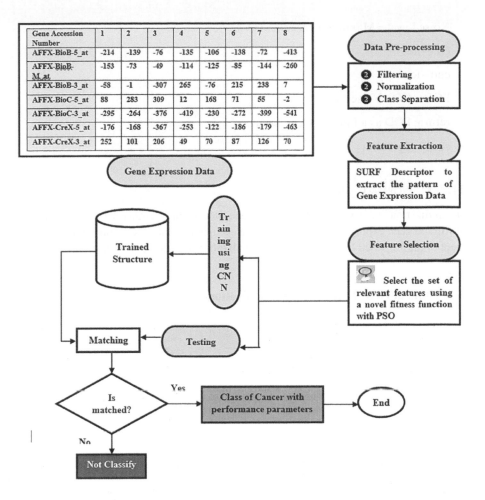

Gene Accession Number	1	2	3	4	5	6	7	8
AFFX-BioB-5_at	-214	-139	-76	-135	-106	-138	-72	-413
AFFX-BioB-M_at	-153	-73	-49	-114	-125	-85	-144	-260
AFFX-BioB-3_at	-58	-1	-307	265	-76	215	238	7
AFFX-BioC-5_at	88	283	309	12	168	71	55	-2
AFFX-BioC-3_at	-295	-264	-376	-419	-230	-272	-399	-541
AFFX-CreX-5_at	-176	-168	-367	-253	-122	-186	-179	-463
AFFX-CreX-3_at	252	101	206	49	70	87	126	70

FIGURE 11.2   Proposed workflow.

Algorithm 1: Gene Expression Filtration

Input: Gene Data (GD) and Estimate noise point (P)

Output: Filtered Gene Data (FGD)

Start filtration

Calculate size of GD, [row, column] = Size (GD)

for m = 1 row

for n = 1 column

Slots = Divide GD into N number of slots

Calculate variation = Check relation between neighbor values of GD

Noise Level = Maximum variation in GD

end – for

end –for

FGD = GD - Noise Level

return: FGD as a Filtered Gene Data

end – Algorithm

11.2.3 Normalization

Normalization is normally utilized to replace the vacant or empty array in the dataset of filtered gene expression data and make a compatible data according to the requirement of the proposed BCC model and normalization algorithm is composed as follows:

Algorithm 2: Gene Expression Normalization

Input: Filtered Gene Data (FGD)

Output: Normalized Gene Data (NGD)

Start normalization

Calculate size of FGD, [Row, Column] = Size (FGD)

For m = 1 row

For n = 1 column

If FGD (m, n) = empty or FGD (m, n) = null

NGD (m, n) = average of FGD

Else

NGD (m, n) = FGD (m, n)

End-if

End – for

End – for

Return: NGD consider as Normalized Gene Data

End – Algorithm

11.2.4 Class Separation

It is a step regarding the separation of the entire dataset according to the blood cancer class like ALL and AML. Based on this step, we can train the BCC model properly that is an effective way to find out the feature pattern of gene expression based on their class and class separation algorithm which are as follows:

Algorithm 3: Gene Expression Class Separation

Input: Normalized Gene Data (NGD)

Output: Class of Gene Data (ALL and AML)

Start separation

Calculate size of NGD, [row, column] = Size (NGD)

for m = 1 row

for n = 1 column

If NGD index belong to 0

ALL (m, n) = NGD (m, n)

else

AML (m, n) = NGD (m, n)

end – if

end – for

end – for

return: ALL and AML consider class of Gene Data

end – Algorithm

11.2.5 FEATURE PATTERN EXTRACTION

After the pre-processing, using SURF descriptor, we need to extract the features or patterns which is based on ALL and AML data. For utilizing or highlighting design extraction because of strength and invariance nature of elements, we use SURF descriptor. The main reason to select the SURF descriptor approach lies in its fast computation behavior of operators using square filtration method to enable the real-time applications such as diseases' classification, pattern recognition, and object tracking. The used algorithm of SURF descriptor is written as:

Algorithm 4: SURF Descriptor for Feature Pattern Extraction

Input: Gene Data (GD)

Output: Feature pattern of Gene Data (F-pattern)

Initialization of feature pattern extraction

Compute the size of GD, [Column, Row] = Size (GD)

for m = 1 row

for n = 1 column

Extrema-detection (m, n) = GD (m, n)

Key-point-localization (m, n) = Extrema-detection (m, n)

If localization need orientation

Orientation (m, n) = Key-point-localization (m, n)

end – if

F-pattern (m, n) = All best Feature

end – for

end – for

return: F- as a feature pattern of gene data either from ALL or AML

end – Algorithm

The information of the descriptor is a picture and a collection of given set of exceptional points selected on a given picture. The final output of the descriptor is the arrangement of component vectors for the unique or original given set of characteristic points. It ought to be noticed that a few descriptors all the while tackle two issues – look for or search for special points and development of descriptors of these points.

The SURF descriptor is a descriptor which simultaneously examines for special points and builds their description, which is invariant to changes in scale and revolution. Furthermore, the searching for various key points in the situation is invariant in the sense that the rotated scene object has a similar set of various feature points as the example or as the sample.

Assurance of exceptional points in the picture or image is performed dependent on the Hessian matrix. Utilizing the Hessian gives pivot or rotational invariance, yet not scale invariance. Thus, SURF applies channels of various scales to ascertain the Hessian.

After applying feature pattern extraction using SURF, The next step is using PSO as an optimization technique for the concept of feature selection. The full details are given in the below sections.

11.2.6 Feature Selection

To accomplish better characterization exactness of BCC model, reduction and choice of component design or features pattern are performed to select a set from the high-dimensional data. There are numerous features or genes present which are irrelevant and do not involve in the training scenario but they increase the chances of error in the model so feature pattern is applied in the BCC model. Feature pattern through selection or reduction approach selects the most effective feature patterns as per the cancer data classes. In this study, we applied the PSO concept with a novel fitness function to select an optimal set of feature patterns. At the same time, it fulfills the goal of feature selection to remove redundant and irrelevant feature pattern for enhancing the BCC model performance. PSO approach of feature selection is considered one of the most substantial steps due to their searching behavior in the machine learning or deep learning field and the used algorithm is written as:

Algorithm 5: PSO as a Feature Selection

Input: F-pattern and Designed FF for feature selection

Output: OF Pattern- Improved feature pattern

First Step: Initialization of Selection

Optimization of F-pattern using Particle Swarm Optimization (PSO)

PSOset up: P – Particle Size based on the F-pattern

V – Velocity of particle

Pp – Particle Position

OF-pattern – Optimized Feature Pattern

Fitness Function:

$F(f) = \{1(\text{True}); \text{ if } F_s * (\text{Pp} + \text{V}) \geq F_t = \text{Threshold}_{\text{Data}} \, 0 \, (\text{False}); \text{Otherwise}$

Where F_s : feature pattern form the F-pattern

F_t : ALL data Threshold and it is the average of all F-pattern

F-pattern Length (Len)

OF-pattern = [] // Set as Null-Empty initially

For I = 1 L

F_s = FS (I) = Selected$_{\text{Data}}$ // Current data from F-pattern

$F_t = \text{Threshold}_{\text{Data}} = \sum_{i=1}^{R} F_s(I)$ // Average of all data (F-pattern)

$$F(f) = Fit \ Fun \ (F_s, F_t)$$

N_{var} = Number of variables

OF-pattern(I) = PSO (F (f), N_{var}, PSO Setup)

end – for

if OF-pattern = 1 then

OF-pattern = Select characteristics form F-pattern

else

OF-pattern = Null

end – **if**

return: OF-pattern improved characteristics of features of the pattern

end – Algorithm

After the feature pattern selection according to the fitness function, we used these features as input of classifier to train the BCC model and here we use the pattern-based CNN as a classifier or deep learning approach. The general description of PSO working is provided below.

PSO

Kennedy and Eberhart [15] stated and also initially brought into the practice that PSO is a heuristic optimization approach. It is at first displayed from swarm knowledge and it relies upon the examination and profound exploration of the group development, i.e., flock movement of the bird and fish. The underlying thoughts on particle swarms were fundamentally pointed toward delivering computational insight or intelligence by utilizing basic analogs of social collaboration, rather than absolutely individual intellectual or cognitive capacities.

The basic idea includes in PSO is that each candidate works in the multi-component search space need to be considered an element or particle having no weight as well as volume, that particles can fly at some pace. The quality of the element or particle is measured through the function called Fitness Function, and the element can be easily modifying the pace w.r.t to the past experience of flying and other individuals. Every particle all times tries to fly the best among other individuals in the group.

If $P_i = (p_{i1}, p_{i2}, ..., p_{in})$ is the current state of particle i, $S_i = (s_{i1}, s_{i2}, ..., s_{in})$ denotes the current speed and $B_i = (b_{i1}, b_{i2}, ..., b_{in})$ corresponds to the best places that all particles have visited. $B_g = (b_{g1}, b_{g2}, ..., b_{gn})$ is the best places that all particles have visited.

11.2.7 MODIFIED CNN

The modified CNN for BCC classification using the gene expression dataset and the genes selected from the previously described subsection were used to optimize the SURF extracted feature pattern. So, CNN is known as modified CNN or MCNN that is designed to classify blood cancer types under different heads such as ALL or AML. This section will exhibit the classification technique proposed in the study to help improve the classification accuracy of the BCC model and also use for system to improve or enhance the performance time. The algorithm of MCNN is as follows:

Algorithm 6: Pattern-Based MCNN

Input: OF Pattern – Optimized feature pattern, Class Category for ALL and AML, Neurons for convey the information

Output: BCC-Pattern qualified pattern along classified output of BCC model.

Begin Training

Initialization the sample-based MCNN with of-pattern: – No of Epochs (E) // Iterations Epochs used by convolutional neural network

–As carrier we use Neuron (N)

– Performance: Data Validation, Error Histogram in the course of the pedagogy and transpose the working characteristic and classes Cross entropy of classes, Gradient.

– Preparing Data Division: Based on Random

Compute span OF-pattern in terms of Len

For j = 1 Len

if OF-Pattern has a place with Type 1 (ALL)

Cat (1) = Feature from the OF-pattern of 1st Part // ALL gene expression data

else (AML)

Cat (2) = Feature from the OF-pattern of 2nd Part // AML gene expression data

end

end

Pattern net initialization using OF-Pattern and Cat

BCC-Structure = Pattern-based MCNN

Training parameter set as per the system requirement and system to be trained accordingly

BCC-Pattern = Training (BCC-Structure, of-pattern, Category (CAT))

Test of Data = sim (BCC-structure, Existing Test Data)

if Testing of Data Group = 1

Categorized Results = ALL with performance evaluation parameters

else

Categorized Results = AML with performance evaluation parameters

return: BCC-Structure as a trained structure with Categorized Results of BCC Model

end

The abovementioned algorithms and procedural steps are the same in both scenarios for BCC model with SURF-based MCNN and BCC model with optimized SURF-based MCNN. The experimental results are briefly described in the following segment of the research article.

Artificial intelligence with deep learning finds an algorithm for solving the original problem on its own, learns from its mistakes, and, after each iteration of training,

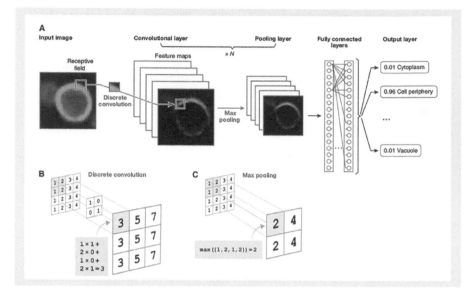

FIGURE 11.3 CNN operation.

gives a more accurate result. Deep learning is used in computer vision (to extract information from images), machine translation, and human speech recognition on audio recordings. Similar to neural network, a convolutional neural network is also composed of collection weights, neurons, and bias values, which are utilized to change the neurons depending on the situation or problem statement. These neurons provide input, calculate the weighted sum, and then carry the data toward the output layer (Figure 11.3).

11.3 EXPERIMENTAL RESULTS

The study proposes a BCC model using PSO-based feature selection approach with SRUF descriptor for MCNN to classify the ALL and AML types of blood cancer. In the proposed model, we present a comparative analysis of classification of ALL and AML blood cancer through gene expression dataset by using SURF-based MCNN and optimized SURF-based MCNN. We also compare the results of simulation of propose model with existing work presented by Chaudhari et al. [16] at the section of this article (Table 11.1).

Figure 11.4 shows the comparison of proposed work using designed BCC model with SURF-based MCNN and BCC with optimized SURF-based MCNN represented by the blue line and the orange line, respectively. From the graph, it has been observed that the values observed for proposed work BSS with optimized SURF-based MCNN is higher than BCC model with SURF-based MCNN. The average precision values examined for BCC model without optimization approach and with optimization approach using MCNN are 0.9682 and 0.9032, respectively. Thus, an improvement of about is 7.2%. Thus, it happened due to the appropriate selection of

TABLE 11.1

Performance Parameters of Proposed BCC Model

No. of Sample	BCC with SURF-Based MCNN				BCC with Optimized SURF-Based MCNN			
	Precision	Recall	F-Measure	Accuracy	Precision	Recall	F-Measure	Accuracy
1	0.814	0.781	0.797	83.53	0.953	0.814	0.878	92.38
2	0.905	0.885	0.894	91.34	0.926	0.938	0.931	95.83
3	0.826	0.772	0.798	80.38	0.968	0.942	0.954	97.93
4	0.913	0.905	0.908	92.74	0.973	0.953	0.962	98.73
5	0.932	0.811	0.867	88.93	0.994	0.963	0.978	99.37
6	0.897	0.875	0.885	89.37	0.974	0.951	0.962	98.94
7	0.878	0.789	0.831	84.94	0.973	0.983	0.977	99.62
8	0.946	0.923	0.934	95.72	0.949	0.918	0.933	96.33
9	0.957	0.919	0.937	95.37	0.978	0.953	0.965	99.73

Source: Authors' research work.

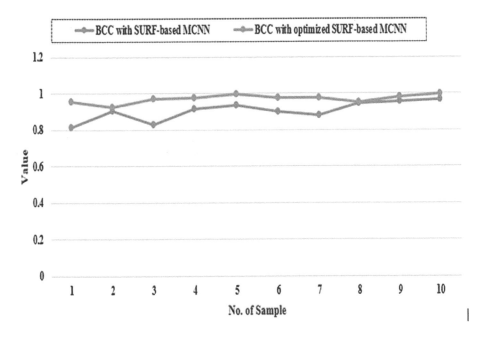

FIGURE 11.4 Comparison of precision.

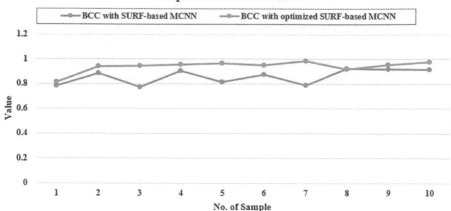

FIGURE 11.5   Comparison of recall rate.

genes features using PSO approach and then increased the accuracy of the trained
BCC model using CNN approach.

The comparison of recall using designed BCC model with SURF-based MCNN
and BCC with optimized SURF-based MCNN for ten samples is shown in Figure 11.5.
Recall represents the number of relevant instances (classification of ALL and AML
type of blood cancer) among the total trained dataset. The observed values for recall
are shown in Figure 11.5. The average recall observed for proposed BCC with optimized SURF-based MCNN and BCC model with SURF-based MCNN is 0.938 and
0.8572, respectively. Thus, an enhancement in the retrieval data from the total available data of 9.43% has been attained.

In the similar way, the comparison of F-measure without optimization and with
optimization approach is shown in Figure 11.6. F-measure parameter is used to show
the balance between the computed precision and recall values of the designed model
in the form of F-measure metrics. The results show better F-measure for the proposed work against the BCC with optimized SURF-based MCNN (Figure 11.7).

As the rate of true positive and true negative to the absolute or total available data
accuracy can be defined. It signifies the overall detection rate of the designed BCC
model. The average accuracy detected using BCC model with SURF-based MCNN
is 89.72%, whereas BCC with optimized SURF-based MCNN is 97.88%. Thus, the
proposed model shows an improvement of 9.09% compared to the BCC model with
SURF-based MCNN (Table 11.2).

PRF and Accuracy Comparison

Figure 11.8 shows the results of proposed a BCC model using the MCNN with
the existing model using the concept of improved K-NN to classify the ALL and
AML types of blood cancer. Thus, the truthfulness of future model is enhanced as
compared to existing model presented by Chaudhari et al. [16] using the concept of
improved K-NN. By utilizing the PSO-based SURF selection of feature pattern in
proposed BCC model, the improvement is achieved and we noted that the improvement is near to 23.42% compared to the existing work.

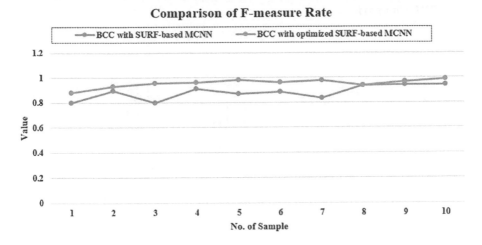

FIGURE 11.6 Comparison of F-measure rate.

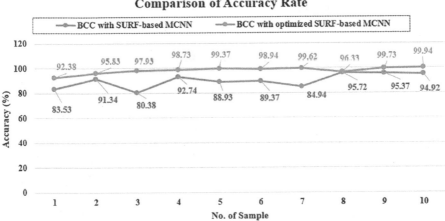

FIGURE 11.7 Comparison of accuracy.

11.4 CONCLUSION

We presented a BCC model using the PSO-based feature selection approach with SRUF descriptor for MCNN to classify the ALL and AML type of blood cancer. The study also shows that feature selection is a very useful approach to optimize the relevant information using SURF descriptor. The feature optimization technique helps to reduce the dataset and hence the accuracy of classifier will be enhanced. The proposed work accuracy was also examined by the comparative study of ALL and AML blood cancer through gene expression dataset in two different manners with SURF-based MCNN and optimized SURF-based MCNN.

(a) PRF Comparison

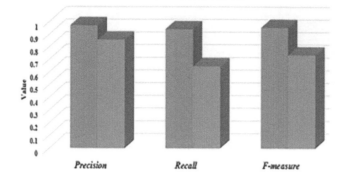

(b) Accuracy Comparison

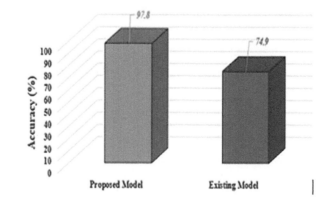

FIGURE 11.8   PRF and accuracy comparison.

TABLE 11.2
Comparative Analysis

Models	Precision	Recall	F-Measure	Accuracy
Proposed Model	0.968	0.938	0.952	97.8
Existing Model [16]	0.855	0.645	0.735	74.9

Source: Authors' research work.

At last, the study investigated the simulation results of proposed BCC model. The test results show better accuracy of about 97.8% and show an improvement of 23.42%. Future scope of the work can be extended by increasing the size of dataset, or by considering the various categories of blood cancer.

REFERENCES

[1] J. Rawat, A. Singh, H. S., & Virmani, J., "Computer aided diagnostic system for detection of leukemia using microscopic images", *Procedia Computer Science*, vol. 70, (2015), pp. 748–756.

[2] S. Agaian, M. Madhukar, & A. T. Chronopoulos, "Automated screening system for acute myelogenous leukemia detection in blood microscopic images", *IEEE Systems Journal*, vol. 8, no. 3, (2014), pp. 995–1004.

[3] M. R. Baer, C. C. Stewart, R. K. Dodge, G. Leget, N. Sulé, K. Mrózek, ... & R. M. Stone, "High frequency of immunophenotype changes in acute myeloid leukemia at relapse: Implications for residual disease detection (Cancer and Leukemia Group B Study 8361)," *Blood, The Journal of the American Society of Hematology*, vol. 97, no. 11, (2001), pp. 3574–3580.

[4] T. L. Dunwell, L. B. Hesson, T. V. Pavlova, V. Zabarovska, V. I. Kashuba, D. Catchpoole, ... & E. R. Zabarovsky, "Epigenetic analysis of childhood acute lymphoblastic leukemia," *Epigenetics*, vol. 4, no. 3, (2009), pp. 185–193.

[5] P. Viswanathan, "Fuzzy C means detection of leukemia based on morphological contour segmentation," *Procedia Computer Science*, vol. 58, (2015), pp. 84–90.

[6] S. Mohapatra, S. S., Samanta, D. Patra, & S. Satpathi, "Fuzzy based blood image segmentation for automated leukemia detection," In *2011 International Conference on Devices and Communications (ICDeCom)* (pp. 1–5). IEEE.

[7] K. H. Chen, K. J. Wang, K. M. Wang, & M. A. Angelia, "Applying particle swarm optimization-based decision tree classifier for cancer classification on gene expression data," *Applied Soft Computing*, vol. 24, (2014), pp. 773–780.

[8] E. Alba, J. Garcia-Nieto, L. Jourdan, &, E. G. Talbi, "Gene selection in cancer classification using PSO/SVM and GA/SVM hybrid algorithms," In *2007 IEEE Congress on Evolutionary Computation* (pp. 284–290). IEEE.

[9] S. Li, X. Wu, & M. Tan, "Gene selection using hybrid particle swarm optimization and genetic algorithm," *Soft Computing*, vol. 12, no. 11, (2008), pp. 1039–1048.

[10] M. S. Mohamad, S. Omatu, S. Deris, & M. Yoshioka, "Particle swarm optimization for gene selection in classifying cancer classes," *Artificial Life and Robotics*, vol. 14, no. 1, (2009), pp. 16–19.

[11] L. Gao, M. Ye, X. Lu, & D. Huang, "Hybrid method based on information gain and support vector machine for gene selection in cancer classification," *Genomics, Proteomics & Bioinformatics*, vol. 15, no. 6, (2017), pp. 389–395.

[12] A. K. Dwivedi, "Artificial neural network model for effective cancer classification using microarray gene expression data," *Neural Computing and Applications*, vol. 29, no. 12, (2018), pp. 1545–1554.

[13] L. Sun, X. Zhang, Y. Qian, J. Xu, & S. Zhang, "Feature selection using neighborhood entropy-based uncertainty measures for gene expression data classification," *Information Sciences*, vol. 502, (2019), pp. 18–41.

[14] T. Golub, D. K. Slonim, P. Tamayo, C. Huard, M. Gaasenbeek, J. P. Mesirov, H., Coller, M. L. Loh, J. Downing, M. A. Caligiuri, C. D. Bloomfeld, E. S. Lander, "Molecular classification of cancer: Class discovery and class prediction by gene expression monitoring," *Science*, vol. 286, (1999), pp. 531–537.

[15] J. Kennedy and R. Eberhart, "Particle swarm optimization," *Proceedings of the IEEE International Conference on Neural Networks*, vol. 4, (1995), pp. 1942–1948. http://dx. doi.org/10.1109/ICNN.1995.488968.

[16] P. Chaudhari, H. Agarwal, & V. Bhateja, "Data augmentation for cancer classification in oncogenomics: An improved KNN based approach," *Evolutionary Intelligence*, vol. 14, no. 2, (2019), pp. 1–10.

12 An Introspective Approach to Fathom Human-Inspired Bipedal Walk Using Gait Analysis for the Matrix of Cyber-Physical System

Manoj Kumar and Pratiksha Gautam
Amity University

Vijay Bhaskar Semwal
NIT Bhopal

CONTENTS

DOI: 10.1201/9781003248750-12

12.1 INTRODUCTION

Bipedal robot [1] walking has been used for more than four decades. Researchers have used numerous computational models to achieve humanoid walking in the robot, yet it is still a challenging task to develop humanoid walking. Gait analysis is the process of analyzing scientifically human walking. It is a combination of continuous as well as discrete dynamics. Biomechanics researchers study motion and ground reaction force to develop humanoid walking [2]. Gait analysis is divided into two phases, stance phase and swing phase, which are further divided into sub-phases. Technology maturity in many aspects has not been reached to develop humanoid walking as a hybrid system. We need high computing devices, hardware, and precision rate to make a proper walk by computer-enabled systems.

12.1.1 HUMAN GAIT ANALYSIS AND KEY TERMS

The process of scientific investigation and study of human walking or human locomotion is called human gait analysis. It has been divided into two sub-phases: stance phase and swing phase. The stance phase is further subdivided into loading response, mid-stance, and terminal stance, while the swing phase has pre-swing, initial swing, mid-swing, and terminal swing. The stance phase is 60% and the swing phase has 40% of the total gait cycle (GC) [3]. Human walk is a combination of many GCs which itself is a combination of continuous and discrete phases. The GC is the period of time of one foot to the next foot of the same side in the same event (Figure 12.1).

12.1.2 HUMANOID MODEL

A variety of models have been developed by domain experts for a few decades. These models have tried to copy human walking, for example, 2D musculoskeletal model,

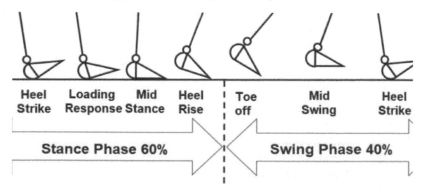

FIGURE 12.1   Gait sub-phases.

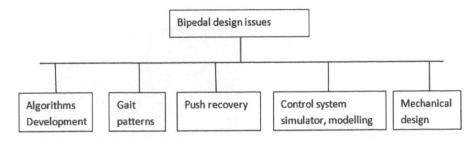

FIGURE 12.2   Different design issues during bipedal development.

2D stick model, deterministic model, 3D gait model, 3D voxel model, hypergraph partitioning model, pose-based model, and data-driven model. Torque is used to generate motion which later generates desired path called a trajectory. A trajectory is calculated online as well as offline, online trajectory is helpful to manage, avoid obstacles, and achieve a predefined path. Zero moment point (ZMP) is a very old concept in the field of robotics that is very useful for a bipedal walk. This concept is further extended and generalized by the name of foot rotation indicator (FRI) [1].

12.1.3   Design, Implementation of Bipeds [4]

Development of bipeds is divided into two categories: hardware and software aspects. Hardware developments deal with mechanical parts (legs, foot, muscle, etc.) and electrical parts (control circuits, circuit simulation, and building of circuit), while software developments correspond to identifying problems, their optimal solutions, design of algorithm, and implementation of it with testing and debugging with finally uploading the program into microcontroller chip of bipeds (Figure 12.2).

12.1.4   Human Walk Vs Bipedal Walk – Limitations and Constraints [5]

Natural walking is governed by a neurological control system; it controls and gives feedback for balance and accuracy in walk-in humans. Human walking has three basic features: (i) human walk completely relies on two legs theory, (ii) legs are fully extended at the time of ground contacts, and (iii) foot strikes with the ground with initial contact of the heel. These features make gait analysis unique in nature from another creature in the world. Different muscles from the thigh, ankle, and knee make the whole walking very supportive and balanced. Bipeds have many inherent limitations like degree of freedom, motor issues, battery energy consumptions, non-linear structure, and controllers' capacities. Generally, the following challenges and constraints lie with bipeds.

Human can walk free in space with different movements. Bipeds have limits in terms of the degree of freedom, muscle, bones, and joint coordination [6]. Human is adaptable to nature, while mostly bipedal robots are still comfortable in a static or controlled atmosphere; it's still a tough task for researchers. Developing smart actuators is a key point in bipedal robots. Bipeds are produced by manufacturers rather than by biology. Geometric configurations like DOF (degree of freedom), link, arms,

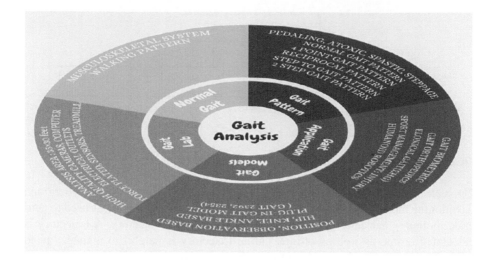

FIGURE 12.3   Knowledge discovery database of gait analysis.

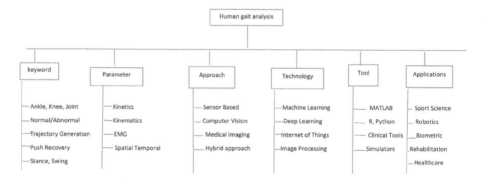

FIGURE 12.4   Conceptual overview of gait analysis framework [8].

and lack of muscle give a complex task for bipedal development. Angular momentum plays a vital role to balance body weight, proper walking, and developing and understanding the human model [7] (Figure 12.3).

12.1.5   HUMAN GAIT ANALYSIS APPROACH [8,9]

This concept came into existence during the 16th century; later on, different concepts were proposed by many researchers from time to time about biomechanics, mainly by Eadweard Muybridge and Leland Stanford, Whittle and Vasconcelos, and Tavares. Human gait analysis has been divided into basically four kinds: using image processing concepts with a camera (vision-based), using the sensor-based approach, using medical imaging-based approach (MRI, CT, X-rays, ultrasound), and other hybrid approaches. Figure 12.4 explains the details about different aspects of human gait analysis like keywords, procedures, approaches, technologies, tools, and applications areas in human gait analysis.

Keywords Used: Human gait uses generally stance, swing, toe on/off, heel on/off, mid-swing, single and double stance and swing phase, GC, ankle, knee, joint, normal/abnormal gait, trajectory, push recovery in its literature [10].

Parameters: Center of pressure (COP), center of mass (COM), force, ground reaction, and muscle response play important roles in gait analysis.

Approaches: To investigate gait, sensors, high-vision cameras, walking posture, occlusions, and data preprocessing techniques play impactful impression.

Technologies: Machine learning, deep learning, Internet of things (IoT), cyber-physical systems (CPS) with hybrid technologies are used to develop the smart robot [3].

Tools Used: Visual gait analysis needs different tools, while quantitative tools are generally R, PYTHON, MATLAB, OpenSim, and WEBOOT.

12.1.6 CHAPTER STRUCTURE

Section 12.1 contains all the fundamental concepts of gait analysis and CPSs. A variety of application areas are discussed in detail in Section 12.2. We use gait analysis as a newly emerging field into CPS in Section 12.3. Research challenges and limitations of work are discussed in Section 12.4. Lastly, Section 12.5 includes the conclusion and future scope of work.

12.1.7 VISION-BASED GAIT ANALYSIS [8,11]

Video camera is used to capture the subjects using frame generations. This approach is of two types: marker-based (direct) and marker-less (indirect). In this, human gait is acquired and preprocessed, and feature selection and feature extraction are performed with the help of suitable image processing techniques, finally, gait classification (healthy/unhealthy) or human identification (male/female) is done with the help of any suitable techniques like machine learning or deep learning.

12.1.8 CPS AND ITS MOTIVATION [12,13]

CPS is a combination of physical components along with software installed on it in a better coordinated manner. This concept is inspired by natural leadership examples that include swarm intelligence and ant colony problem, where different agents work internally or externally, in a distributed manner or centralized way based on dynamic conditions. CPS clubs many technologies like the IoT, artificial intelligence (AI), robotics, VLSI (very large-scale integration), and nanotechnology. Capabilities of CPS have given rise to the concept of human-computer interface (HCI), smart robotics, and industrial automation versions.

With all possible considerations, CPS has produced many challenges such as cyber threats, M2M (machine-to-machine) interface, self-configurations, self-managements, and self-optimization in connection with adapting diverse cultural challenges, protection of edge computing devices from external hacking, and applying modeling and simulations. It's really interesting to know how all together work as machines into cyberspace to do better will decision-making support and cost-effective processes. Through this work, we have tried to focus on how gait analysis for humanoids is changing the mindset of CPS for future generations.

12.2 APPLICATIONS OF GAIT ANALYSIS

12.2.1 Sport Science [14,15]

Athletes' performances and injury prevention can be monitored and training cum outcome is being analyzed for faster recovery of players. Foot pressure and motion parameters measurements are taken into count to improve the quality of players. Gait assessment is very helpful in developing injury-free running or practice; it benefits sports persons and runners who cannot afford sports coaches or gait experts or who have a poor running approach. Assessment and treatment-based systems are the future of this domain using AI.

12.2.2 Physical Rehabilitation [16]

This is the reason to perform clinical gait analysis and to achieve better results; there is a significant need to identify and perform a better selection mechanism among multiple available treatments. Diagnosis, assessment, monitoring, and prediction are the key to rehabilitation. Biomechanical measurements in terms of accuracy, validation, cost-effectiveness, reliable treatments, and procedures should be developed and the highest standard practices in clinical practice should be followed by patients. Further research should be done to validate an old and new range of optimized techniques and only best suited models over to conventional models should become into practice for patients. Joint movements, joint angle, and soft tissue displacements could be developed using advanced machine learning and AI approaches for better estimation of foot movements to produce better gait of humans. Skin-mounted sensors, cameras, and advanced lab equipment related to gait can produce better training and principles in this domain.

12.2.3 Clinical Applications

Clinical gait assessment and gait research are two main aspects of gait analysis. Both lead to improving medical treatments and making qualitative decisions related to human healthy walk. Data reliability, suitability, and repeatability in performing clinical practice should be taken highly precise and corrective in manner. Data related to hip, knee, ankle, and pelvis position should be normally distributed in experiments to get accurate and error-free results [16,17]. Quantitative gait analysis is based on high-vision cameras, quality sensors, and better labs to record data to make better analyses and decisions. Kinematics, kinetics, analysis of muscle activity, and energy level are being interpreted by different available tools of gait and better decisions are being taken in the clinical field. Gait abnormalities, disease identification-related to the brain, paralysis, and Parkinson's disease are a few names where gait analysis plays a major role [18].

12.2.4 BIOMETRIC APPLICATIONS [19,20]

Video images and radar systems are generally two types of approaches used for people identification. Join, hip rotations, and GCs are being extracted and generated patterns are being matched against other samples in order to identify the particular subject. Gait as a new biometric could be utilized for criminal activity and surveillance applications, people identification, and for developing a more secure hybrid system using traditional biometric features like voice, eyes, face, and fingerprints. It suffers from feature selection, consistency, quality dataset availability, and current technological paradigm impact to get more secure and reliable belief.

12.2.5 PROSTHETIC LIMBS DESIGN [21,22]

There are numerous materials and technologies used to design artificial limbs. Designing the joints and knee limbs with similar reproduction of subjects like actions and walk, functions are very complicated as it needs to understand kinetics, kinetic, efficiency, and structural shape and size measurements. Gait analysis provides qualitative and quantitative measurements and a variety of data to fill the gap between original and artificial actions in subjects [22]. User needs are being accessed with the help of rehabilitation specialists and physicians. Initial alignments and custom designing are made and then the final fitting is done. Prosthetic feet and ankle mechanism, finding limbs curve, and developing an automatic system to improve walking is new dimensions of this work [21,22].

12.2.6 AR/VR APPLICATIONS [23]

1. Gait lab equipped with high-vision cameras and body sensors are using augmented reality-based systems. Markers, orientation, and position in the human body to get quality data in the motion system have been increased in automatic labs. A normal, virtual, and augmented version of walk comparisons parameters [24] shows that it's quite difficult to maintain a stable position during walk-in treadmill, paper results like the center of mass, the center of pressure, stride length, and other parameters of gait suggest that people have walked unusual way in a treadmill which affects rehabilitation treatments of person. This discussion-based research paper results indicate that head-mounted AR/VR devices will provide better feedback to walk in treadmills or to walk in terrain for a humanoid robot. Its future lies in the analysis of position control in gait biomechanics based on statistics data recorded from AR/VR labs. AR/VR automated labs are in trend nowadays for gait training to monitor fatigue, anxiety, depression, falling rate, and quality of life.

12.2.7 Humanoid Robotics

Bipedal robot walk is inspired by a human walk. Human gait analysis is based on the statistical data of hip, Knee, ankle position, and orientation to training robots to walk much like human beings. Sophia, ASIMO, Atlas, Aquanaut, etc. are just a few names of humanoid robots. Trajectory generation, optimization of trajectory, smooth walk, obstacle avoidance walks, jerk-free walk, walk-in controlled – uncontrolled environments, and monitoring daily activities of humans are just a few concepts that are taken from human gait analysis. Robot training [25] and path planning are very complex and time-consuming processes. The development of suitable algorithms for a robot for walking, very much similar to humans, is still in its early stage. Nowadays, deep learning, machine learning, natural language processing, and sentiment analysis are being used progressively to implement humanoid robot from computer science prospects. Humanoid robots have been used to teach students, serve tea, take care of elders, and send them to war and remote areas. Due to inherent limitations and challenges, it's really difficult to replicate the human mind, body, and walk. Future scope lies in jerks free walk, minimum energy consumptions, obstacle avoidance, medical sector, agriculture sector, morphological computation aspects, and development of cognitive approaches in connection with interaction with environment dynamically.

12.2.8 Human Activity Recognition

Human gait is used to identify the person and classify whether a person is having a normal walk or abnormal walk. The latest research [26] is working on finding the emotions of the walker as people have different walking based on their mood, emotions. Slow walk, medium walk, and fast walk can be classified in the case of gait analysis which further helps to get insight into the elderly patient's smooth walk and neurological disease identifications. The future application of gait analysis is to develop automatic emotion recognizers, developing assistance for patients and in the biometric field.

12.2.9 Expert Systems

Automated gait analysis can be used for decision-making in healthcare sectors and sports injuries. These systems can easily suggest proper walking posture in treadmills, terrains, and out of laboratories. Impaired gait and muscles responsible [27] for dysfunction gait pattern development should be automated and a data-driven approach should be used to extract useful information which will help to make decisions in case of normal and abnormal walking. Data analysis expert system and gait pathological expert system must be improved with rapid changes in technology.

12.2.10 Orthopedic Care [28]

Gait analysis is performed either visually or with the help of specialized instruments. Instrumented gait analysis has gained popularity as a complete and standard analysis due to capturing joint angle movements, joint force recording, and muscle activity along with neurological examinations. Children and patients with cerebral palsy disease diagnosis come under this category. Normal and abnormal walk, types of impaired gait identifications, and their useful recommendations come from orthopedic care.

12.3 GAIT ANALYSIS NEW EMERGING SECURITY DIMENSIONS IN CPS

With the rapid growth in data and computational devices in networks, the importance of CPSs has gained an eye to researchers and academicians' community. We need high computation, strong networking skills, and control management to interact with physical processes. Gait analysis has been stated to develop human-machine interaction with the IoT uses. Gait generation and design of suitable controllers for a bipedal walk have gained tremendous popularity in humanoid fields. Humanoid robots in the field of hotels, restaurants, assistantship for people, housework like cleaning, lifting heavy work, interaction with visitors, teaching and training purposes, and industry applications fields have gained popularity in the last few decades. During all work, proper controlling, feedback mechanism, and remote controlling in the network have made this more difficult. A robot is a good example of a CPS that can be statically or dynamically controlled, where instructions given remotely are being monitored. Fall risk evaluation, BP calculation, human activity recognition, occlusion, and gait as a new security layer to existing bioinformatics systems need storing instruction in the cloud and communicating signals to system side, high-vision cameras, large computation facility, and better algorithms. Due to the increasing number of technologies, the heterogeneous device uses, and inherent limitations of robotics, this domain has a lot of limitations and challenges [29].

12.4 RESEARCH CHALLENGES AND LIMITATIONS

- Trajectory optimization with/without a jerk, minimum energy consumption, and obstacle avoidance in a remote manner
- Teaching and training robots are very cost-effective and complex processes due to human-machine interaction due to regulate direct command and govern
- Modeling and framing human motion into bipeds are still challenging due to limitations of sensors data acquisitions properly, noise and filter process computation
- Development of traditional and non-traditional controllers to execute generated gait
- Privacy as a malicious attack, instruction detection, safety constraints, scalability and complexity managements, feedback systems involvement, etc. are gait inherent limitations that make gait more secure for general use
- Gait as an additional layer into existing layers like voice recognition, thumb detection can be used to make a more secure CPS
- Development of feasible computer algorithms to manage and control humanoid in uncontrolled environments like search operation in coal mines, remote mining, and search and rescue operations during the war

- IoT and CPS make the robotic field more complex and exploitations of vulnerabilities demand to make it more securable and less risk-effective in nature

12.5 CONCLUSIONS AND FUTURE SCOPE

Gait analysis as a CPS has shown promising applications in the area of precision farming, remote mining, search and rescue, nanorobotics for healthcare fields, and the driverless car. Developing smart robot-like human beings needs multidimensional knowledge, high computation power, physical resources, and better cloud computing concepts. Apart from diverse applications areas, gait analysis has opened challenged for robotics and automation society to rethink and reshape it more strongly as CPS for future. Gait as CPS has complex and diverse applications and challenges and their solutions differs case to case. Future scope lies on the development of cultural, social and green computing-based robots which are quick to take decision and actions with edge analysis very near to AI. CPS for future could be autonomous robot, natural language processing (NLP)-like Chabot, voice assistants, and Human in Loop.

REFERENCES

[1] Kljuno, E., and R. L. Williams. "Humanoid walking robot: modeling, inverse dynamics, and gain scheduling control." *Journal of Robotics* 2010, Article ID 278597 (2010): 19. https://doi.org/10.1155/2010/278597.
[2] Sinnet, R. W., M. J. Powell, R. P. Shah, and A. D. Ames. "A human-inspired hybrid control approach to bipedal robotic walking." *IFAC Proceedings Volumes* 44, no. 1 (2011): 6904–6911.
[3] Akhtaruzzaman, Md, A. A. Shafie, and Md R. Khan. "Representation of human gait trajectory through temporospatial image modelling." *ARPN Journal of Engineering and Applied Sciences, Asian Research Publishing Network (ARPN)* 11, no. 6 (2016): 4105–4110.
[4] Al-Shuka, H. F. N, F. Allmendinger, B. Corves, and W.-H. Zhu. "Modeling, stability and walking pattern generators of biped robots: a review." *Robotica* 32, no. 6 (2014): 907.
[5] Capaday, C. "The special nature of human walking and its neural control." *Trends in Neurosciences* 25, no. 7 (2002): 370–376.
[6] Fukui, T., Y. Ueda, and F. Kamijo. "Ankle, knee, and hip joint contribution to body support during gait." *Journal of Physical Therapy Science* 28, no. 10 (2016): 2834–2837.
[7] Herr, H., and M. Popovic. "Angular momentum in human walking." *Journal of Experimental Biology* 211, no. 4 (2008): 467–481.
[8] Prakash, C., R. Kumar, and N. Mittal. "Recent developments in human gait research: Parameters, approaches, applications, machine learning techniques, datasets and challenges." *Artificial Intelligence Review* 49, no. 1 (2018): 1–40.
[9] Kavanagh, J. J., and H. B. Menz. "Accelerometry: A technique for quantifying movement patterns during walking." *Gait & Posture* 28, no. 1 (2008): 1–15.
[10] Abbass, S. J., and G. Abdulrahman. "Kinematic analysis of human gait cycle." *Al-Nahrain Journal for Engineering Sciences*, 16, no. 2 (2013): 208–222.
[11] Singh, J. P., S. Jain, S. Arora, and U. P. Singh. "Vision-based gait recognition: A survey." *IEEE Access* 6 (2018): 70497–70527.

[12] https://www.cognizant.com/whitepapers/how-cyber-physical-systems-are-reshaping-the-robotics-landscapecodex4484.pdf.

[13] Yaacoub, J. P. A., Salman, O., Noura, H. N., Kaaniche, N., Chehab, A., & Malli, M. Cyber-physical systems security: Limitations, issues and future trends. *Microprocessors and Microsystems*, 77, (2020): 103201.

[14] Wahab, Y., and Bakar, N. A. Gait analysis measurement for sport application based on ultrasonic system. In *2011 IEEE 15th International Symposium on Consumer Electronics (ISCE)* (pp. 20–24). IEEE, (2011).

[15] http://www.gaituk.com/assessments/injury-prevention-performance-enhancement-gait-assessment/.

[16] Baker, R. Gait analysis methods in rehabilitation. *Journal of Neuroengineering and Rehabilitation*, 3, no. 1 (2006): 1–10.

[17] Benedetti, M. G., Catani, F., Leardini, A., Pignotti, E., & Giannini, S. (1998). Data management in gait analysis for clinical applications. *Clinical Biomechanics*, 13, no. 3, 204–215.

[18] https://musculoskeletalkey.com/applications-of-gait-analysis.

[19] https://witanworld.com/article/2021/06/09/future-of-biometrics-gait-biometrics/.

[20] Connor, P., & Ross, A. Biometric recognition by gait: A survey of modalities and features. *Computer Vision and Image Understanding*, 167, (2018), 1–27.

[21] Esquenazi, A. Gait analysis in lower-limb amputation and prosthetic rehabilitation. *Physical Medicine and Rehabilitation Clinics of North America*, 25, no. 1, (2014), 153–167.

[22] Daniele, B. Evolution of prosthetic feet and design based on gait analysis data. In Ernesto Iadanza (ed.), *Clinical Engineering Handbook* (pp. 458–468) (2020). Academic Press: United Kingdom.

[23] Nagymáté, G., & Kiss, R. M. Affordable gait analysis using augmented reality markers. *PLoS One*, 14, no. 2, (2019): e0212319.

[24] Chan, Z. Y., MacPhail, A. J., Au, I. P., Zhang, J. H., Lam, B. M., Ferber, R., & Cheung, R. T. Walking with head-mounted virtual and augmented reality devices: effects on position control and gait biomechanics. *PLoS One*, 14, no. 12, (2019): e0225972.

[25] Li, M., & Wang, D. Physical motion gait analysis and planning for humanoid robots. *International Journal of Simulation Systems, Science & Technology*, 17, no. 32 (2016): 48.1–48.8. doi: 10.5013/IJSSST.a.17.32.48

[26] Xu, S., Fang, J., Hu, X., Ngai, E., Guo, Y., Leung, V., … & Hu, B. *Emotion Recognition from Gait Analyses: Current Research and Future Directions.* arXiv preprint arXiv:2003.11461. (2020).

[27] Bontrager, E. L., Perry, J., Bogey, R., Gronley, J., Barnes, L., Bekey, G., & Kim, J. W. GAIT-ER-AID: An expert system for analysis of gait with automatic intelligent preprocessing of data. In *Proceedings of the Annual Symposium on Computer Application in Medical Care* (p. 625), (1990). American Medical Informatics Association: USA.

[28] White, H., & Augsburger, S. Gait Evaluation for Patients with Cerebral Palsy. In: Nowicki, P. (ed) *Orthopedic Care of Patients with Cerebral Palsy*, Springer, Cham., Switzerland, (2020): 51–76. https://doi.org/10.1007/978-3-030-46574-2_4

[29] Sagar, A. K., L. Banda, S. Sahana, K. Singh, and B. Kumar Singh. "Optimizing quality of service for sensor enabled Internet of healthcare systems." *Neuroscience Informatics*, 1, no. 3 (2021): 100010. ISSN: 2772-5286, https://doi.org/10.1016/j.neuri.2021.100010.

13 Capacitated Vehicle Routing Problem Using Algebraic Particle Swarm Optimization with Simulated Annealing Algorithm

Mohammad Sajid
Aligarh Muslim University

Jagendra Singh
Bennett University

Ranjit Rajak
Dr. Harisingh Gour Central University

CONTENTS

DOI: 10.1201/9781003248750-13

13.1 INTRODUCTION

In intelligent logistics, the efficient distribution of demands is the key to competent services, which is essential in realizing smart cities [1]. Vehicle routing problem (VRP), a well-recognized combinatorial optimization problem, was introduced by George Dantzig and John Ramser in 1959 [2]. It has multiple applications in the industry as it is of enormous importance for numerous organizations that deliver services to consumers [3]. The VRP's objective is to make total traveled distance minimum while satisfying the capacity, flow, and integrity constraints [3,4]. The capacitated VRP (CVRP) is a well-recognized NP-hard and combinatorial optimization problem and several exact methods, heuristics, local search operators, and evolutionary algorithms have been employed to search for the near-optimal solutions acceptable computational time [3,5–8]. The CVRP is inherently discrete, i.e., the solutions are intrinsically discrete in nature and can be represented as permutations (discrete solutions) of the given customers. There are two categories of evolutionary and swarm optimization algorithms to address the discrete problems: continuous-valued and discrete-valued. In discrete-valued evolutionary algorithms, the solutions are represented as permutations or integer-valued arrays, and the decoding method is not required. In the case of continuous-valued evolutionary algorithms, the solutions are represented as a continuous-valued array. A decoding method is essential to convert a continuous-valued array into a discrete solution that describes the solution of the problem under consideration. Various decoding methods exist, such as random-key (RK), reverse RK, and angle modulation, to convert a continuous-valued array into a discrete solution [9–11]. The issue with the decoding methods is that infinite continuous-valued vectors can be decoded to a sole discrete solution because of cardinality; thus, continuous-valued algorithms may need to explore bulky plateaus in the fitness space. However, numerous continuous-valued (numerical) evolutionary algorithms were applied to resolve the discrete optimization problems [12–15].

Particle swarm optimization (PSO), established by Eberhart and Kennedy [16], starts with particles consisting of continuous values, and then the particles are evolved over generations to search for the global best particle with the best possible objective value. PSOs' particles move in the solution space according to the mathematical equations of position and velocity. The PSO can employ the decoding method to convert a continuous-valued particle into a solution [9–11]. These methods have different techniques to convert a particle into an integer permutation for CVRP and other permutation-based problems. Due to cardinality, an infinite number of continuous-valued solutions can be mapped to a single permutation, leading to wastage of computational efforts. It may also lose the perception of basic PSO. Thus, the conversion methods also require extra computational efforts that could have been saved if solutions remained permuted [12].

Many works employ different algorithms to address the CVRP and its variants in literature using PSO, simulated annealing (SA), genetic algorithm, gravitational emulation local search and genetic algorithm, and firefly algorithm [4,7,17,18,19–23].

No algebraic algorithm exists to solve the CVRP; however, algebraic algorithms have been proposed for other permutation problems [12–15]. This work investigates an algebraic version of PSO to solve the CVRP, which works based on permutation-based solutions. The algebraic PSO is combined with SA algorithm to avoid trapping in local optima and premature convergence [15,24–26]. Thus, algebraic PSO with SA (APS) algorithm is developed to resolve CVRP to make the total distance traveled minimum in order to supply all customers' demands. The solutions of the CVRP can be represented by the set of permutations of given customers, and this set forms a symmetric group (a.k.a. permutation group) with a composition operator. The proposed APS algorithm employs algebraic operators defined in [12], and local search operators [24].

The arrangement of the remaining chapter is as follows. The CVRPs' mathematical formulation is presented in Section 13.2, while Section 13.3 discusses the permutation group preliminaries. The proposed APS algorithm is explained in Section 13.4, while the implementation and simulation study are given in Section 13.5. Section 13.6 puts the final remarks along with the potential research directions.

13.2 PROBLEM FORMULATION

In the CVRP [3,4], the aim is to determine the routes with the least total distance for multiple homogeneous vehicles visiting a set of customers to fulfill their demands. The CVRP can be explained in terms of the Euclidean graph $G = (N, E)$, where N is the set of nodes (vertices) of the graph and is given by $N = \{0, 1, 2, 3......n\}$, and $E = \{(i, j) : i, j \in N, i \neq j\}$ is a set of edges connecting the nodes. The main depot is characterized by node 0, and the customers are denoted by nodes 1, 2, 3...n. For each node $i \in N$, the demand q_i and geometrical coordinates (x_i, y_i) are given. The travel distance d_{ij} connecting two nodes $i, j \in N$ can be computed with the help of their geometric coordinates. The main depot has zero demand, while the demands of the customers can be represented by $Q = \{q_i : 1 \leq i \leq n\}$. To satisfy the demands of all customers, a fleet of K homogeneous vehicles is given and represented by $M = \{m_k : 1 \leq k \leq K\}$ and the capacity of each vehicle is assumed to be W. The objective is to supply the needs of all customers using the available vehicles to make the total distance traveled minimum while observing the constraints of flow, capacity, and integrity. Let X_{ij}^k is the decision variable whose value can be either 0 or 1. The values of the decision variable are as follows:

$$X_{ij}^k = [1 \text{ If vehicle } m_k \text{ moves from node } i \text{ to node } j \quad 0 \text{ Otherwise} \quad (13.1)$$

The CVRP can be mathematically expressed as:

$$\min F = \sum_{m_k \int M} \sum_{i \int N} \sum_{j \int N} X_{ij}^k * d_{ij} \tag{13.2}$$

with the following constraints

$$\sum_{m_k \in M} \sum_{j \in V} X_{ij}^k = 1, \quad \forall i \in N \tag{13.3}$$

$$\sum_{m_k \in M} \sum_{i \in N} X_{ij}^k = 1, \quad \forall j \in N \tag{13.4}$$

$$\sum_{i \in N} \sum_{j \in N} X_{ij}^k * q_i \leq W, \quad \forall m_k \in M \tag{13.5}$$

$$\sum_{j \in N} X_{0j}^k = 1, \quad \forall m_k \in M \tag{13.6}$$

$$\sum_{i \in N} X_{i0}^k = 1, \quad \forall m_k \in M \tag{13.7}$$

$$\sum_{i \in N} X_{ih}^k - \sum_{j \in N} X_{hj}^k = 0, \quad \forall\, h \in N - \{0\}, i \neq j, i \neq h, j \neq h \quad \forall m_k \in M \tag{13.8}$$

$$X_{ij}^k \, \varepsilon \, \{0, 1\}, \quad \forall i, j \in N, \forall m_k \in M \tag{13.9}$$

Equation (13.2) shows the objective function, which computes the total distance traveled by all K vehicles to serve n customers. As per constraints (13.3) and (13.4), every customer must be served only once using one vehicle only. As per eq. (13.5), the customers' demands assigned to a vehicle must be equal to or below the vehicles' capacity W. Equations (13.6) and (13.7) ensure the vehicles' flow constraint according to which every vehicle must begin and end the journey at the main depot. Equations (13.8) and (13.9) maintain vehicles' flow constraints at every customer and integrity of the decision variable X_{ij}^k, respectively.

13.3 PERMUTATION GROUP PRELIMINARIES

CVRP is a well-acknowledged combinatorial NP-hard optimization problem in which permutations of integers (or n customers) are the solutions from the set of all permutations. Figure 13.1 shows two solutions (permutations) α and β for the CVRP consisting of nine customers. If α and β are solutions of the CVRP, there exists composition operator $*$ so as $\alpha*\beta = \alpha(\beta(i))$, $\forall\, i \in V$, is again a solution. Figure 13.2 shows the resultant permutation γ of the composition of two permutations α and β given in Figure 13.1. In this work, we employ the group theory to simulate the moves of the bare-bones PSO algorithm to resolve the considered problem. This section discusses the group theory preliminaries corresponding to the permutation group.

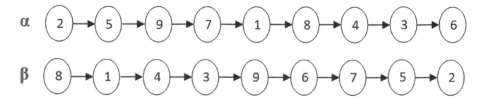

FIGURE 13.1   Two permutations for the CVRP with nine customers.

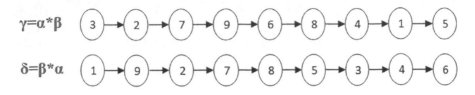

FIGURE 13.2 Resultant permutations after applying the composition in two different ways.

13.3.1 PERMUTATION GROUP

A set of permutations P is said to be a group if it contains any two permutations α and β; and a binary composition operation * such that $\alpha*\beta$ is again a permutation which is an element of P. The permutation group also satisfies the conditions of identity, and inverse elements as well as associativity [12]. There exist an identity element $e \in P$, and inverse element $\alpha^{-1} \in P$ for each permutation group.

The permutation group satisfies three conditions of the group. If permutation group P consists of more than two permutations, it is a non-abelian group, i.e., $\alpha*\beta \neq \beta*\alpha$ for all $\alpha, \beta \in P$. The permutation group P is a finite group if it contains a finite number of elements, i.e., it consists of $n!$ permutations for a set N of n elements (customers). The permutation group P is also a symmetric group because each permutation is the bijection of itself. A subset $\langle P \rangle$ of P exists, which is a group under the composition operator (*) and $\langle P \rangle$ is known as the generator group if all elements of P can be generated using the elements of $\langle P \rangle$ and their inverses. The group P and the generating set $\langle P \rangle$ can be characterized by a connected directed graph known as the Caylay graph. The Cayley graph $\Gamma = (\langle P \rangle, P)$ is a directed graph with the following properties.

- Every element $\alpha \in P$ is a node in the graph.
- Every $\rho \in \langle P \rangle$ is assigned a color c_ρ.
- For each $\alpha \in P$ and $\rho \in \langle P \rangle$, there exists a directed edge of color c_ρ to the node $\alpha*\rho \in P$.

13.3.2 ABSTRACT ALGEBRAIC OPERATIONS

In numerical evolutionary algorithms like PSO and DE, the operations are applied on vectors to generate a new vector representing the new solution for the problem under consideration. The numerical evolutionary algorithms require decoding methods to convert the continuous-value vector into permutation, representing the solution. For combinatorial permutation spaces, it is necessary to define addition, subtraction, and scalar multiplication operators such that the resultant solution remains a permutation. For combinatorial permutation space, it has been proved in the research works [12–15] that the abstract addition ($\oplus$), abstract subtraction ($\ominus$), and abstract scalar multiplication ($\odot$) can be defined in terms of composition operator (*) such that these operations hold the same properties analogous to the numerical operations in Euclidean vector space.

As proved in [12], the algebraic abstract addition ($\oplus$) and subtraction ($\ominus$) operators are independent of the generators and generating set given as

$$\gamma = \alpha \oplus \beta = \alpha * \beta \qquad \forall \alpha, \beta \in P \qquad (13.10)$$

$$\gamma = \alpha \ominus \beta = \beta^{-1} * \alpha \qquad \forall \alpha, \beta \in P \qquad (13.11)$$

The β^{-1} represents the inverse permutation which is defined as

$$\beta^{-1} = \left(k_i, k_j, k_l \ldots\right) \text{ s.t., } \beta(k_i) < \beta(k_j) < \beta(k_l) \ \forall k_i \text{ are indices of } \beta \qquad (13.12)$$

It is to be noted that abstract addition and subtraction operators are both non-commutative due to the non-commutative property of composition operator (*), i.e., $\alpha \oplus \beta \neq \beta \oplus \alpha$, and $\alpha \ominus \beta \neq \beta \ominus \alpha$.

To define the abstract scalar multiplication, it is required to choose a generating set forming the symmetric permutation group P. The set consisting of all permutations that swap the elements of adjacent positions is known as the set of all simple transpositions $\langle P_{ST} \rangle$. The generating set $\langle P_{ST} \rangle$ consists of $n-1$ elements, and its search space diameter is $\binom{n}{2}$. In this work, the generating set of all simple transpositions $\langle P_{ST} \rangle$ has been chosen due to its simple search space structure, i.e., the composition of $\alpha \in P$ and $\rho_i \in \langle PST \rangle$ results in the swapping of elements at indices i and $i+1$ in α. The successive application of this simple property is helpful to transforming permutation $\rho_i \in \langle PST \rangle$ into the desired permutation.

The analogy for abstract scalar multiplication is similar to scalar numerical multiplication in Euclidean vector space. The multiplication of a scalar φ with permutation $\alpha \in P$ changes the length of permutation α as per the value of scalar φ. For given $\varphi \geq 0$ and $\alpha \in P$, the abstract scalar multiplication ($\varphi \odot \alpha$) fulfills the following three conditions to simulate the scalar numerical multiplication in Euclidean vector space [12].

C1: $\alpha \odot \varphi = \left[\alpha.|\varphi|\right]$; **C2**: $\alpha \odot \varphi \leq \alpha \quad \forall \varphi \in [0,1]$; **C3**: $\alpha \odot \varphi \geq \alpha \quad \forall \varphi > 1$

There may be more than one permutation in P due to applications of C1–C3 and non-unique minimal decompositions of each permutation. The diameter also restricts the value of scalar φ, i.e., the scalar value must be a fraction of diameter. Due to the above restrictions, abstract scalar multiplication can be realized in many different ways. To implement the abstract scalar operation ($\odot$) on permutations, the following four cases are considered based on properties C1–C3.

$$\alpha \odot \varphi = e \text{ if } \varphi = 0 \qquad (13.13)$$

$$\alpha \odot \varphi = \alpha \text{ if } \varphi = 1 \qquad (13.14)$$

$$\alpha \odot \varphi \leq \alpha \text{ if } 0 < \varphi < 1 \qquad (13.15)$$

$$\alpha \odot \varphi \geq \alpha \text{ if } \varphi > 1 \qquad (13.16)$$

Equations (13.13–13.16) follow the analogy of scalar numerical multiplication in Euclidean vector space. The abstract scalar multiplication can be computed using eqs. (13.13 and 13.14) and stochastic bubble sort algorithm [12–15].

13.4 THE PROPOSED ALGORITHM

PSO algorithm is encouraged by the collective swarms' intelligence, and it is capable of solving complex optimization problems [16]. Every particle is characterized by position and velocity, which are continuous-valued vectors. The following equations are employed to update the kth particle's position and velocity.

$$v_k^{t+1} = \omega * v_k^t + c_1 * r \text{ and } {}^{\{0,1\}} * \left(zb_k - z_k^t\right) + c_2 * r \text{ and } {}^{\{0,1\}} * \left(zb^t - z_k^t\right) \tag{13.17}$$

$$z_k^{t+1} = z_k^t + v_k^{t+1} \tag{13.18}$$

where v_k^t and v_k^{t+1} are kth particles' velocities in iteration t and $t+1$, respectively; z_k^t and z_k^{t+1} are the kth particles' positions in iteration t and $t+1$, respectively; and zb^t and zb_k are the position of the best particle at iteration t and the best position of the kth particle so far, respectively. The r and ${}^{\{0,1\}}$ characterizes a random number in the [0,1], and c_1 and c_2 correspond to the learning factors, while ω denotes the inertia weight.

The goal of the PSO is to find out a vector $z_k^t = \left[z_1 z_2 z_3 \ldots z_v\right]$ that optimizes the fitness function $f\left(z_k^t\right)$ of the problem under consideration. The dimensions of the position vector depend on the dimensions of the problem. In PSO, position and velocity vectors consist of continuous values for discrete and continuous optimization problems. For CVRP, the position vectors must convert into permutations using RK, angle modulation, or other methods.

13.4.1 INITIAL POPULATION AND FITNESS VALUE

For APS, the population of size η is initialized with random permutations of given customers in the CVRP. Figure 13.3a shows the random particle (permutation) consisting of nine customers.

The particle α can be decomposed into routes using the cluster-first and route-second method. This method assigns customers to the first vehicle in the same order as they appear in the particle until the capacity is exhausted. Next, the remaining particles' remaining customers are allocated to the next vehicle until the second vehicles' capacity is exhausted. This method is employed to decide routes for all customers.

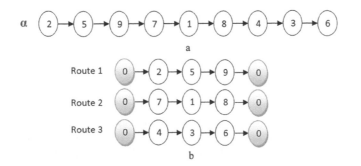

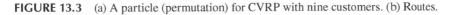

FIGURE 13.3   (a) A particle (permutation) for CVRP with nine customers. (b) Routes.

Figure 13.3b depicts the routes generated from the particle given in Figure 13.3a by applying the cluster-first and route-second method.

13.4.2 The Proposed Algorithm

Unlike PSO, APS does not use any decoder function, and it simulates the moves of the PSO algorithm in permutation search space. The pseudo-code of the APS algorithm is given in the text-box. The APS algorithm requires the CVRP instance and a total number of iterations θ. The APS algorithm fetches the CVRP instance to determine the customers, their geometric coordinates, and demands. Next, the positions and velocities of η particles in the population are initialized to random permutations of total customers in the CVRP instance. To produce the population for iteration $t+1$, every particle's velocity and position are boosted.

Proposed APS Algorithm

Input: CVRP instance, #Iterations (θ)

Output: Best Particle

Initialize position of η particles with random permutations

Initialize the velocity of η particles with random permutations

Initialize personal best positions of particles by their initial positions

Initialize learning factors c_1 and c_2

Initialize inertia weight ω

While ($\theta > 0$)

For each particle,

Use eq. (19) to update the velocity

Use eq. (20) to update the position

      Compute fitness value of each particle

Mark global best particle

Apply SA on global best particle

Return best particle

Let $Z_k^t = [Z_1 Z_2 Z_3 \ldots \ldots Z_v]$ and V_k^t be the kth particles' position and velocity in iteration t of the APS algorithm. The position and velocity of a particle are expressed as permutations of given customers in the CVRP instance. Assume ZB^t and ZB_k are the best particle of population and kth particles' best position so far, respectively. The kth particles' velocity is computed based on the following components

$$I_1^{t+1} = \left(\omega \odot V_k^t \right) \tag{13.19}$$

$$I_2^{t+1} = c_1 \times r \text{ and } (0, 1) \odot \left(ZB_k \ominus Z_k^t \right) \tag{13.20}$$

$$I_3^{t+1} = c_2 \times r \text{ and } (0, 1) \odot \left(ZB^t \ominus Z_k^t \right) \tag{13.21}$$

Now, the velocity of kth particle for iteration $t+1$ is computed as follows:

$$V_k^{t+1} = I_1^{t+1} \oplus I_2^{t+1} \oplus I_3^{t+1} \tag{13.22}$$

To compute the velocity, the highest precedence is given to abstract scalar multiplication ($\odot$), and the abstract addition ($\oplus$) and abstract subtraction ($\ominus$) have the same priority. For these three operators, associativity is considered from left to right.

The position of the kth particle for iteration $t+1$ is computed as follows:

$$Z_k^{t+1} = Z_k^t \oplus V_k^{t+1} \tag{13.23}$$

Similarly, the remaining particles of the population for iteration $t+1$ are computed. Next, the fitness value of each particle is calculated using eq. (13.2) and cluster-first and route-second approach as discussed in Section 13.5.1.

Simulate annealing improves the convergence speed and avoids trapping in the local optimum, based on local search operators, i.e., 2-opt*, exchange, and relocate operators [24–26]. In the beginning, SA requires the best particle ZB^{t+1} of the population at $t+1$ iteration and then searches the neighborhoods of the given best particle. Initially, the temperatures' value τ is initialized to a constant value. Next, the neighborhood particles (solutions) are searched by the stochastic application of 2-opt*, exchange, and relocate operators. An operator is selected randomly, and a neighborhood solution is generated. The better-quality neighborhood solution is accepted; otherwise, the worst solution with some probability value based on Boltzmann's constant is accepted. It is obtained as follows:

$$\text{Prob}_A = \{1 \quad \Delta C \le 0 \, e^{-\frac{\Delta C}{\theta}} \quad \Delta C > 0 \tag{13.24}$$

$$\tau = \tau * (1 - \varnothing) \tag{13.25}$$

where the ΔC represents the change in objective values and $\varnothing$ is the rate of temperature change.

All steps of the loop are repeated for a fixed iteration. At last, the APS algorithm returns the global best particle. Every particle saves its own best positions and adjusts the velocity considering its own best performance and best performing particle. The particles are using their personal best, socially best particle followed by the application of SA to search the neighborhood of the given particle. Eventually, the swarm will converge to near-optimal positions with good convergence speed.

13.5 SIMULATION STUDY

To compare the performance, the proposed APS and classical RK-PSO algorithms have been implemented in Python 3.9.6 on DESKTOP-L9O54PJ, Intel(R) Core ™

TABLE 13.1

Parameters' Values

S. No.	Parameter	Values
1	Swarm size (η) and generations (θ)	200
2	Learning coefficient (c_1) and (c_2)	0.01
3	Inertia weight (ω)	0.09
4	Experiments conducted for each CVRP instance	05
5	Objective value recorded	Best, mean, standard deviation

i7–8700, Windows-10. The RK-PSO algorithm employs classical settings. Each particle is the vector of continuous values between −10 and 10, and total values in the vector are equal to total customers in the CVRP instance. To compute the fitness value (total traveled distance), the RK-PSO algorithm uses the RK method to convert the continuous-valued vectors into permutation followed by the computation of total traveled distance [9–11]. Table 13.1 presents the parameters' values that are used to realize the proposed APS and RK-PSO algorithms. For APS and PSO algorithms, the swarm size (η), generations (θ), inertia weight (ω), acceleration coefficients c_1 and c_2 are 200, 200, 0.09, 0.01, and 0.01, respectively. For each CVRP instance, the algorithms are executed five times, and the mean values, standard deviations of total distance traveled, and the best particle is recorded.

Augerat et al. proposed dataset A consisting of 27 CVRP instances [27]. In each instance, the vehicle capacity is 100 due to its homogeneous nature, while the total customers, customers' demand and locations, and depots' location are distinct. The total customers vary between 32 and 80; the total vehicles vary between 5 and 10; the demands vary between 1 and 72; and the distances vary between 2 and 125.87.

13.5.1 SIMULATION RESULTS

The algorithms were tested for 27 CVRP instances, but it is not possible to show the graphical results for all the instances. Therefore, we provide the visual results for only three instances. Figures 13.4, 13.5, and 13.6 show the results for the instances A-n32-k5, A-n44-k6, and A-n60-k9, respectively. Figures 13.4a, 13.5a, and 13.6(a) represent the routes generated by the APS algorithm corresponding to the A-n32-k5, A-n44-k6, and A-n60-k9 CVRP instances, respectively. Figures 13.4b, 13.5b, and 13.6b show the routes generated by the RK-PSO algorithm for the A-n32-k5, A-n44-k6, and A-n60-k9 CVRP instances, respectively. It is clearly visible that the routes' overlapping is minimal in the routes offered by the APS algorithm compared to the RK-PSO algorithm. The minimum overlap indicates that the customers in the same region are served by minimum possible vehicles. Thus, minimum overlapping directly affects the total traveled distance by the vehicles. For the A-n32-k5 instance, APS and RK-PSO offered the total traveled distances as 797.87 and 1542.55; mean as 835.59 and 1561.88 for five experiments; standard deviation as 16.81 and 16.94, respectively. For the A-n44-k6 instance, APS and RK-PSO offered the total traveled distances as 959.37 and 1881.88; mean as 1023.90 and 1942.28; standard deviation as

45.15 and 41.75, respectively. For the A-n60-k9 instance, APS and RK-PSO offered the total traveled distances as 1463.44 and 2887.80; mean as 1524.88 and 2942.07 for five experiments; standard deviation as 38.49 and 51.30, respectively. As given, the APS algorithm's standard deviations are slightly more compared to those shown by the RK-PSO algorithm. However, the mean values provided by the APS algorithm are significantly less than the mean values offered by the RK-PSO algorithm. Thus, it indicates the superiority of the proposed APS algorithm.

Figures 13.4c, 13.5c, and 13.6c show the total traveled distances generated by the APS and RK-PSO algorithms over 200 generations corresponding to the A-n32-k5, A-n44-k6, and A-n60-k9 CVRP instances, respectively. It can be analyzed that the APS algorithm offered the best optimized total traveled distances compared to the RK-PSO algorithm.

Table 13.2 presents the results of all 27 CVRP instances of dataset A. The table includes the best distances traveled, means, and standard deviations for the best particles in the five experiments APS and RK-PSO algorithms offer. As we can check in Tables 13.2 for all instances of datasets A, the solutions produced by the APS

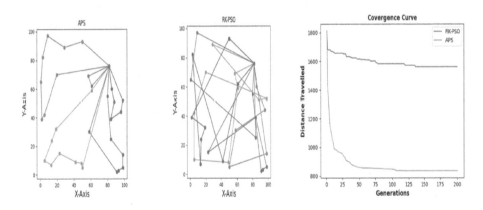

FIGURE 13.4   Results for A-n32-k5.

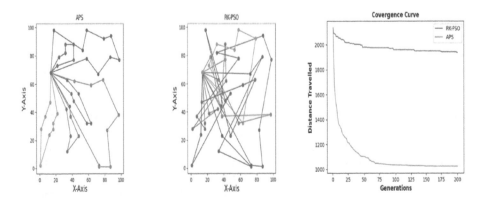

FIGURE 13.5   Results for A-n44-k6.

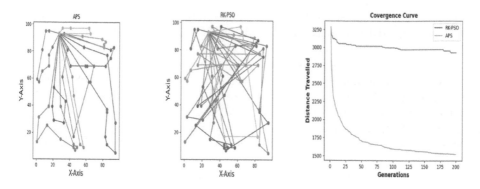

FIGURE 13.6   Results for A-n60-k9.

TABLE 13.2
The Outputs for CVRPs' Instances of Set A

CVRP Instances	RK-PSO Algorithm			APS Algorithm		
	Best Solution	Mean	Standard Deviation	Best Solution	Mean	Standard Deviation
A-n32-k5	1542.55	1561.88	16.81	797.87	835.59	16.94
A-n33-k5	1327.44	1368.16	58.07	676.10	703.06	19.90
A-n33-k6	1268.25	1291.33	31.15	748.16	798.34	37.05
A-n34-k5	1268.25	1291.33	31.15	748.16	798.34	37.05
A-n36-k5	1538.08	1570.56	25.89	816.41	868.56	29.86
A-n37-k5	1426.51	1488.91	36.81	716.03	756.79	31.70
A-n37-k6	1638.55	1679.49	33.69	973.08	987.66	22.01
A-n38-k5	1533.75	1561.64	25.99	751.09	769.81	17.14
A-n39-k5	1631.66	1688.84	60.37	866.07	908.82	25.83
A-n39-k6	1790.73	1803.00	20.62	840.06	892.45	47.28
A-n44-k6	1881.88	1942.28	45.15	959.37	1023.90	41.75
A-n45-k6	2185.40	2213.43	27.53	1013.78	1088.16	48.23
A-n45-k7	2051.36	2095.58	36.65	1180.92	1251.19	43.93
A-n46-k7	1970.67	2061.20	57.06	951.80	1043.90	52.71
A-n48-k7	2274.94	2299.98	23.53	1114.60	1196.73	62.81
A-n53-k7	2350.31	2432.53	56.31	1115.76	1142.68	27.56
A-n54-k7	2540.22	2605.71	71.15	1302.73	1327.06	17.86
A-n55-k9	2448.91	2647.04	296.88	1131.43	1197.49	46.17
A-n60-k9	2887.80	2942.07	42.11	1463.44	1524.88	36.75
A-n61-k9	2443.68	2495.54	38.49	1105.69	1163.89	51.30
A-n62-k8	3002.26	3097.71	59.82	1449.61	1509.93	43.55
A-n63-k9	3373.24	3423.73	35.83	1791.85	1837.51	52.58
A-n63-k10	2892.52	2962.84	82.51	1436.45	1536.42	67.65
A-n64-k9	2896.45	2991.85	56.15	1512.87	1602.77	64.10
A-n65-k9	3042.09	3121.64	55.66	1329.51	1364.78	32.86
A-n69-k9	3133.76	3163.56	31.76	1280.45	1352.47	53.09
A-n80-k10	4041.47	4158.35	91.34	2004.43	2072.19	56.58

algorithm are significantly better than the solutions produced by the RK-PSO algorithm in terms of total traveled distances, means, and standard deviations.

The followings are the observations of the simulation study:

- For all instances of set A, the proposed APS algorithm offered the best values of total traveled distances compared to the RK-PSO algorithm. The mean values of total traveled distances for five experiments provided by the proposed APS algorithm are significantly lower than the RK-PSO algorithm's mean values. For some instances, the standard deviation values provided by the APS algorithm are slightly higher than the standard deviation values offered by the RK-PSO algorithm. Thus, the proposed APS algorithm performs better than the RK-PSO algorithm for all CVRP instances of set A.
- As can be seen that the overlapping of routes is minimal in the routes generated by the APS algorithm in comparison to the routes generated by the RK-PSO algorithm. The minimum overlap indicates that the customers in the same region are served by minimum possible vehicles. Thus, minimum overlapping directly affects the total traveled distance by the vehicles. This is one of the reasons for APS algorithms' superior performance.
- There are two reasons for the performance improvement of the APS algorithm. First, the APS consists of particles in the form of permutations only rather than the continuous-valued vectors. The proposed APS algorithm generates the particles using algebraic operators, and thus, these operators help to evolve over different generations and also help to save the computational efforts of the algorithm. Second, the SA algorithm searches the best neighborhood particles using local search operators (2-opt*, relocate, and exchange). SA algorithm helps to minimize the overlapping between routes and improve the quality of particles.
- The proposed APS algorithm may also be used to resolve the variants of CVRP, and other permutation-based optimization problems. To extend the proposed algorithm, reinforcement learning algorithms cloud environment can be used [28–33].

13.6 CONCLUSION

In this work, APS algorithm is investigated to resolve an NP-hard and well-recognized CVRP to make total traveled distance minimum. The proposed APS algorithm combines the algebraic version of PSO and SA algorithms to explore the solution space for global solutions and exploit a given solution's neighborhood, respectively. The APS algorithm uses permutations as solutions and employs an algebraic composition operator to realize the different operations of PSO. The continuous-valued moves of continuous solution space are imitated in the discrete solution space using an algebraic composition operator on permutations. The well-known 27 CVRP benchmark instances have been evaluated to measure the performance of the APS and RK-PSO algorithms. The simulation study clearly shows that APS performs consistently better compared to RK-PSO.

The proposed APS algorithm can also be extended to resolve the variants of CVRP. It may also be developed to solve other permutation-based optimization problems such as drone-routing problems, independent-task scheduling problems, dependent tasks scheduling problems, and virtual machine scheduling problems.

REFERENCES

[1] X. Liu, Y. X. Liu, N. N. Xiong, N. Zhang, A. F. Liu, H. L. Shen, C. Q. Huang, "Construction of large-scale, low-cost delivery infrastructure using vehicular networks," *IEEE Access*, 6, 21482–21497, 2018.

[2] G. Dantzig, J. H. Ramser. "The truck dispatching problem." *Management Science*, 6, 80–91, 1959.

[3] P. Toth, D. Vigo, *"Vehicle Routing: Problems, Methods, and Applications"* Philadelphia, PA: Society for Industrial and Applied Mathematics, 2nd Ed., 2015.

[4] M. Sajid, A. Zafar, S. Sharma, "Hybrid genetic and simulated annealing algorithm for capacitated vehicle routing problem," *2020 6th IEEE International Conference on Parallel, Distributed and Grid Computing (PDGC), JUIT Waknaghat, India*, 131–136, 2020.

[5] J. Y. Potvin, "State-of-the art review – Evolutionary algorithms for vehicle routing," *INFORMS Journal on Computing*, 21, 518–548, 2009.

[6] R. Baldacci, A. Mingozzi, R. Roberti, "Recent exact algorithms for solving the vehicle routing problem under capacity and time window constraints," *European Journal of Operational Research*, 218(1), 1–6, 2012.

[7] M. Sajid, J. Singh, R.A. Haidri, M. Prasad, V. Varadarajan, K. Kotecha, D. Garg, "A novel algorithm for capacitated vehicle routing problem for smart cities," *Symmetry*, 13(10), 1923, 2021.

[8] J.F. Cordeau, M. Gendreau, G. Laporte, J.Y. Potvin, F. Semet, "A guide to vehicle routing heuristic," *Journal of the Operational Research Society*, 53, 512–522, 2002.

[9] M. Ayodele, J.A.W. McCall, O.R. Coudert, "RK-EDA: A novel random key-based estimation of distribution algorithm," in *Proceedings of 14th International Conference on Parallel Problem Solving from Nature, PPSN XIV*, 849–858, Edinburgh, UK, 2016.

[10] L. Liu, W. Liu, D.A. Cartes, I.Y. Chung, "Slow coherency and angle modulated particle swarm optimization-based islanding of large-scale power systems," *Advanced Engineering Informatics*, 23(1), 45–56, 2009.

[11] B.J. Leonard, A.P. Engelbrecht, "Angle modulated particle swarm variants," in *Proceedings of 9th International Conference on Swarm Intelligence, ANTS 2014, LNCS*, 38–49, 2014.

[12] V. Santucci, M. Baioletti, A. Milani, "An algebraic framework for swarm and evolutionary algorithms in combinatorial optimization," *Swarm and Evolutionary Computation*, 55, 100673, 2020.

[13] V. Santucci, M. Baioletti, G.D. Bari, "An improved memetic algebraic differential evolution for solving the multidimensional two-way number partitioning problem," *Expert Systems with Applications*, 178, 2021.

[14] V. Santucci, M. Baioletti, A. Milani, "Algebraic differential evolution algorithm for the permutation flowshop scheduling problem with total flowtime criterion," *IEEE Transactions on Evolutionary Computation*, 20(5), 682–694, 2016.

[15] M. Baioletti, A. Milani, V. Santucci, "Algebraic particle swarm optimization for the permutations search space," in *Proceedings of 2017 IEEE Congress on Evolutionary Computation, CEC 2017, Donostia, Spain*, 1587–1594, 2017.

[16] J. Kennedy, R. Eberhart, "Particle swarm optimization," In *Proceedings of IEEE International Conference on Neural Networks*, vol 4, 1942–1948, 1995.

[17] Y. Marinakis, M. Marinaki, A. Migdalas, "Particle swarm optimization for the vehicle routing problem: A survey and a comparative analysis" In Martí R., Panos P., Resende M. (eds.) *Handbook of Heuristics*. Springer, Cham.

[18] S.M. Raza, M. Sajid, J. Singh, "Vehicle routing problem using reinforcement learning: Recent advancements," *3rd International Conference on Machine Intelligence and Signal Processing, NIT, Arunachal Pradesh, 2021, LNEE*, vol 858, 269–280, 2022.

[19] A.A.R. Hossainabadi, A. Slowik, M. Sadeghilalimi, M. Farokhzad, M.B. Shareh, A.K. Sangaiah, "An ameliorative hybrid algorithm for solving the capacitated vehicle routing problem," *IEEE Access*, 7, 175454–175465, 2020.

[20] İ. Ilhan, "A population-based simulated annealing algorithm for capacitated vehicle routing problem," *Turkish Journal of Electrical Engineering & Computer Sciences*, 28(3), 1217–1235, 2020.

[21] N. Lin, Y. Shi, T. Zhang, X. Wang, "An effective order-aware hybrid genetic algorithm for capacitated vehicle routing problems in internet of things," *IEEE Access*, 86102–86114, 2019.

[22] T. Azad, M. Ahsan Akhtar Hasin, "Capacitated vehicle routing problem using genetic algorithm: A case of cement distribution," *International Journal of Logistics Systems and Management*, 32(1), 132–146, 2019.

[23] M. Sajid, H. Mittal, S. Pare, M. Prasad, Routing and scheduling optimization for UAV assisted delivery system: A hybrid approach, *Applied Soft Computing Journal*, 126, 109225, 2022

[24] L. İlhan, "An improved simulated annealing algorithm with crossover operator for capacitated vehicle routing problem," *Swarm and Evolutionary Computation*, 64, 100911, 2021.

[25] Y. Xiao, Q. Zhao, I. Kaku, N. Mladenovic, "Variable neighborhood simulated annealing algorithm for capacitated vehicle routing problems," *Engineering Optimization*, 46(4), 562–579, 2014.

[26] P. Chen, H.K. Huang, X.Y. Dong, "Iterated variable neighborhood descent algorithm for the capacitated vehicle routing problem," *Expert Systems with Applications*, 37(2), 1620–1627, 2010.

[27] Augerat et al., *The VRP Web*. [Online]. Available: https://neo.lcc.uma.es/vrp/vrp-instances/capacitated-vrp-instances/, 2013.

[28] R. A. Haidri, M Alam, M. Shahid, S. Prakash, M. Sajid, "A deadline aware load balancing strategy for cloud computing," *Concurrency and Computation: Practice and Experience*, 34(1), e6494, 2021.

[29] M. Sajid, Z. Raza, "Energy-efficient quantum-inspired stochastic Q-HypE algorithm for batch-of-stochastic-tasks on heterogeneous DVFS-enabled processors," *Concurrency Computation: Practice and Experience*, 31(20), e5327, 2019.

[30] M. Sajid, Z. Raza, "Energy-aware stochastic scheduler for batch of precedence-constrained jobs on heterogeneous computing system," *Energy*, 125, 258–274, 2017.

[31] M. Sajid, Z. Raza, M. Shahid, "Hybrid bio-inspired scheduling algorithms for batch of tasks (BoT) applications on heterogeneous computing system," *International Journal of Bio-Inspired Computation*, 11(3), 135–148, 2018.

[32] S. Sharma, M. Sajid, "Integrated fog and cloud computing: issues and challenges," *International Journal of Cloud Applications and Computing (IGI)*, 11(4), Article 10, 2021.

[33] S. Sahana, K. Singh, "Fuzzy based energy efficient underwater routing protocol," *Journal of Discrete Mathematical Sciences and Cryptography*, 22(8), 1501–1515, 2019.

14 An Innovative Smart IoT Device to Measure and Monitor Patient's Critical Parameters in Hospitals

*Naveen Rathee, Jaishanker Prasad Keshari,
and Sandeep Kumar*
IIMT College of Engineering

Varnika Rathee
VIT Bhopal

Gaurav Sinha
IIMT College of Engineering

Imran Ahmed Khan
Jamia Millia Islamia

Monika Singh
Brahmanand Mahavidhyalaya

CONTENTS

DOI: 10.1201/9781003248750-14

241

14.1 INTRODUCTION

In the current scenario, when a person gets under critical illness, immediate hospitalisation becomes necessary for seeking medical aid and in maximum cases, they are being transferred to ICU directly. In ICU, the doctors want to continuously measure vital parameters to see the current condition and after seeing the test reports of the patient they go for further treatment. During this testing and monitoring, the patient has to go through another hard time during their monitoring by the healthcare technician as there are different cables attached to the body for measuring (blood pressure, heart rate, ECG, etc.) on one side and CPU along with monitor on the other side. This makes the patient very uncomfortable as his movements are restricted because any movement of the body can affect the precision and accuracy of the parameters being monitored. While the patient going to the uncomforted zone can lead to other issues and moreover in cases of some patients (Kids, Senior Citizens) it sometimes leads to stress and other side effects.

The application of IoT in the field of healthcare and biomedical plays a vital role in almost all its domains and thereby serving society and humanity at large. The word IoT is framed by combining two words i.e. Internet of Things (IoT). Internet refers to the interconnection of millions of computers and IoT devices. With the help of the internet, users can access all the data based on some standard protocol, while things are concerned with connectivity over the device at any place.

14.1.1 IoT Implementation in Transforming Business

The unpredictable electrical energy situation, in the midst of India's developing power interest and proficiency, is upgraded by the reception of advancement in IoT in Indian dispersion organisations. At the point when energy is incorporated with the web, then, at that point it is defined as energy Internet. This innovation assists with gathering ongoing data on power creation, appropriation and transmission. IoT likewise upholds sustainable power joining into the framework in a productive manner. IoT gives two procedures to making this framework a self-ruling block chain and Software-characterised organising. The two innovations have their upsides and downsides, which are examined. The issues identified with interoperability, solid correspondence and network safety in Indian utilities are additionally tended to [1,2].

14.1.2 Giant Energy Management Employing IoT Technology

An organisation of actual frameworks, associated with one another through the web, is known as the IoT. These actual frameworks may be any input gadget with sensors and linked programming. Input gadgets are associated with the remote organisation for intercommunication with other gadgets. They are used for observing, finding, investigating numerous exercises. All such gadgets need to utilise a similar web convention (IP) that is associated with the web.

Intelligent frameworks are the core parameters of IoT. The IoT gadget should have great dependability for keeping up with energy productivity; secured framework:

The information from a gadget is given to the cloud and afterward to other passageways, either through a wired or remote framework. This information can be inclined to security dangers. Henceforth, embrace appropriate wellbeing measures for information security like gadget verification, firewall, IPS (Intrusion anticipation framework), block chain, SDN (programming characterised network) and so on. Supportability of the energy age and effective appropriation of power is conceivable due to impromptu observing and oversight of energy utilisation, utilising very good quality IoT instruments, for example, information mining [2,3].

14.1.3 IoT a Boost in Smart Farming

The significance of IoT and information examination has been introduced for powerful and productive cultivating rehearses. The wireless sensor network (WSN) hubs and their association with the web are planned and introduced. The IoT framework has a splendid future in the agribusiness area as it is the ideal counterpart for it. For better simplicity of combination and augmentation of framework utilisation, the public authority should overhaul their approach. The primary constraint is cost and framework information. The public authority ought to acknowledge the significance of the data that can be created from the IoT framework and consequently support ranchers for the equivalent giving less expensive advances and straightforward entry to IoT gadgets and administrations. There is expanded exploration of IoT framework which will give ideal IoT arrangements in the farming area [1].

This Innovation involves utilisation of IoT, efficient intelligent sensors, for transmitting the critical measured parameters from multiple sensors to an ARM Cortex-M3 high-precision microcontroller along with a wireless transmitter circuit that can be attached to or can be placed in her/his pocket. This information received is transmitted wirelessly to a desktop computer receiver in the ICU zone, which is further networked wirelessly to the healthcare staff room. If any abnormal readings are observed, an alert to the staff desktop or on their mobiles is sent. The same alert also triggers the patient relative for sensing any undo conditions. This Innovation removes almost 80% of the cables currently used and increases the ease of comfort for a patient in the ICU in the hospital. The main objective of this smart IoT device is the smart and compact wireless remote monitoring features for blood pressure, pulse rate, SpO2, EEG, rate of respiration, temperature and level of blood glucose with the help of intelligent sensors and systems.

Integrating various sensors to get measured parameters and then processing them using advanced computing algorithms are incorporated in the embedded controller. ARM controller with signal conditioning unit for extraction and filtering noise signal is attached, so that the large volume data available from the sensors can be transmitted. An advanced computing algorithm is developed to de-noise the signal and further data compression technique is utilised making the device more efficient and reliable. A unique ID, battery-operated, rechargeable device will be assigned to each patient with miniaturisation soft and limited electrode connections in contrast to the existing conventional systems. With the availability of free frequency bands with the existing medical gadgets, the same is incorporated in our system also without interference to other existing devices. The relatives of the patient accompanying are also aimed for in this approach near the ICU waiting lounge area. Evaluation and validation are fully incorporated in viewing the results.

This work eliminated the traditional approach to health care by incorporating IoT-enabled devices so that the quality of health services can be improved.

The scope of this research can be summarised as follows:

1. An IoT-enabled gadget can collect real-time information regarding the patient's status.
2. After retrieving the information about the patient, the data can be classified and processed further in real time.
3. This IoT-enabled device provides a solution to the patient's critical parameters at anywhere and anytime.

Thus, innovative smart IoT-enabled device continuously monitors ICU patient critical parameters thereby removing the necessity of lesser healthcare staff. It further removes the discrepancy caused due to human errors. There occurs efficient communication and sharing of information between doctors, nurses and patient relatives for the effectiveness of health service at large with this IoT-enabled device. Moreover, this real data is uploaded to the cloud to further process using the interconnection of network and smart devices.

The remainder of the study is organised as follows: Section 14.2 discusses the literature survey and paper contribution. Section 14.3 presents an overview of the proposed system. Section 14.4 shows the results and discussion of IoT in healthcare monitoring system. Section 14.5 shows conclusion and future scope followed by references at the end of this study.

14.2   LITERATURE SURVEY

The author stressed the application of IoT devices for continuously monitoring the health parameters of the patient in article [4]. The use of IoT devices for smart home automation and healthcare industry has been depicted in article [5,6]. The design framework system for healthcare integrated with IoT devices has been implemented in article [7]. The use of cloud computing and web service for the design of an e-healthcare system is focussed in article [8]. Further design of health monitoring systems using IoT-based devices is incorporated in [9,10]. Detailed analysis and its use of IoT for measuring healthcare and agriculture parameters are discussed [11,12].

Internet of Medical Things (IoMT)-based health application is designed to meet the challenges of security and privacy of smart devices [13]. In articles [14,15], the research work focussed on opportunities and challenges in IoT healthcare system design. Further, the design of a reliable IoT-based healthcare system for measuring various parameters of a patient has been implemented in article [16]. The author implemented the design of an edge-computing–based biomedical IoT-enabled device [17]. The health parameters can be gathered by making the use of algorithm and framework based on wearable devices to further process the data in real time is implemented in article [18], while the use of big data has been designed and implemented for healthcare in article [19,20].

The author has implemented an emergency system of the IoMT Smart Hospitals using ICT tools and usages in article [21]. The author has designed and implemented an IoT-based smart home [22,23]. The importance of IoT in agriculture, farming along with the green house in the context of developing nations has been stressed in papers [1,2,24,25].

14.3 PROPOSED SYSTEM

The basic component of IoT is sensors integrated with actuators and microcontrollers device. This smart IoT device has made the medical equipment intelligent enough to capture the patient's data in real time so that the doctors can easily analyse the patient's data within a fraction of time. This measured patient data can be sent to the cloud for clinical diagnosis from anywhere at any place.

The entire system is designed in three steps:

Step 1: To design a portable system using a wireless interface between the IoT compact system, mobiles and desktops, provide alertness to healthcare staff along with patients' relatives. Hardware using an ARM CortexM3-based processor along with a wireless transmitter is embedded to connect different sensors through thin wires and place it on the patient's bed or in the vicinity of his body. The hardware is operated using a battery that is rechargeable. It is a portable device with soft connections to the sensors in comparison to the existing conventional system in hospitals. Because of the freely available frequency bands for portable medical gadgets, using them in the ICU ward doesn't create any interference with other tools. Management of individual IoT devices for patients in the ICU is ensured. Memory management slot in each module is ensured for faithful storage and restructured of data as and when required for analysis is taken care of.

Step 2: A novel signal processing algorithm is designed for the analysis and processing of the data arriving from sensors and is wirelessly shared to desktop or other required devices. The algorithm has two levels of processing the signals: the first is for the removal of noise and compressing data; the second is for enhancement and marking required for the smart display that is to be implemented on the desktop of patients' ICU ward. This received data from sensors is sampled easily using ARM CortexM3 single processor. The third level, data analysis and alert activation, will be implemented in the clinical team room's desktop computer that will also acquire information from other rooms. This prominent feature will assist in alerting personnel if something goes wrong with the patient.

Step 3: Testing the design novel efficient compact system in the laboratory and evaluating and testing the system remotely using various efficient smart medical sensors, the results obtained were very promising and satisfactory.

14.3.1 PROMINENT OUTCOMES OF THE PRODUCT

This innovation prompts the accompanying results:

1. Design and advancement of a smaller IoT gadget framework that incorporates different remote sensors to survey and screen different basic and essential boundaries in patients (principally recuperating patients). Wireless receiver connected to many detectors on a desktop within patient's ICU room, controller desktop inside the clinical team area connected to a desktop in the ICU ward room. Healthcare staff mobile wirelessly connected to the desktop in the ICU ward room available with each patient.
2. Wirelessly deliver the computed sensor signals to the desktop, notifying workers if the values are abnormal.
3. Developed new and efficient algorithms for evaluating sensor signals.
4. Evaluated and then validated the compact system design.
5. Remote interface between the ARM controller system, IoT device and a desktop/computer is developed.
6. Created a network between the user's desktop and the hospital's desktop in order to send an alarm signal.

Figure 14.1 shows a block diagram of a smart compact IoT-embedded system, while Figure 14.2 shows the proposed architecture of the innovative smart IoT device for measuring and monitoring critical parameters in hospitals and Figure 14.3 shows

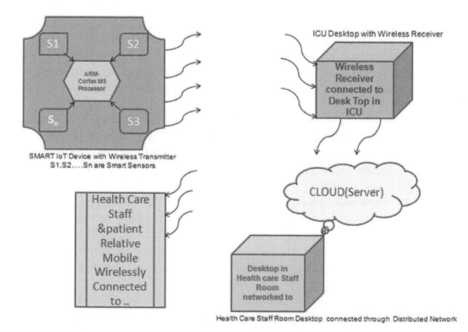

FIGURE 14.1   Block diagram of smart compact IoT-embedded system.

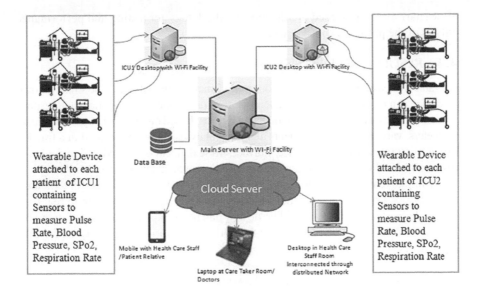

FIGURE 14.2 Proposed architecture of the innovative smart IoT device for measuring & monitoring critical parameters in hospitals.

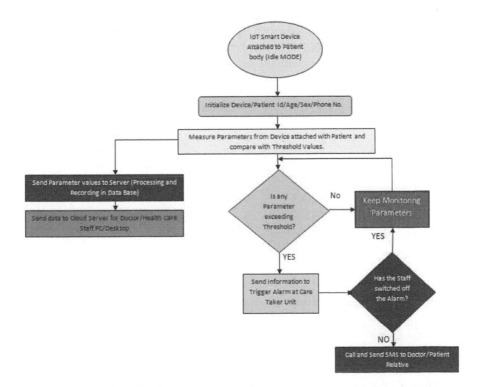

FIGURE 14.3 Flow chart for the proposed architecture methodology.

a flow chart for the proposed architecture methodology for testing and performance analysis of the proposed smart device.

14.4   RESULT AND DISCUSSION

IoT is able to measure and monitor various body functions in real time. The smart innovative device can record heart rate data, temperature data, blood pressure data and respiratory rate data, respectively, as shown in below figures. Based on the need of patient critical parameters, various test measurements and data analyses can be performed in a fraction of time using this IoT-enabled device. Further, these data can be stored on big data platforms which can be accessed by any medical professional at any place and anytime.

The implementation of smart innovative IoT devices is extremely useful to healthcare staff, doctors, patients and their relatives since the data can be seen in real time with the help of a cloud platform. The design has been implemented based on a prototype model; hence, all the pathological health parameters can be analysed for easy diagnosis and correction.

14.4.1   TESTING AND PERFORMANCE ANALYSIS

Figure 14.4 highlights the performance analysis of the system designed for bpm with an error rate with the existing system. This is based on various patients' temperature data collected by actual machine available in ICU and observed data measured by designed smart IoT-proposed device. It also measures their percentage error as recorded in hospital. Figure 14.5 shows the performance analysis of a system designed for body temperature with the error rate of the existing system. Figure 14.6 measures actual and observed data along with the percentage error of blood pressure of various patients as observed. All data are taken at Malik Hospital, Panipat, Haryana, India.

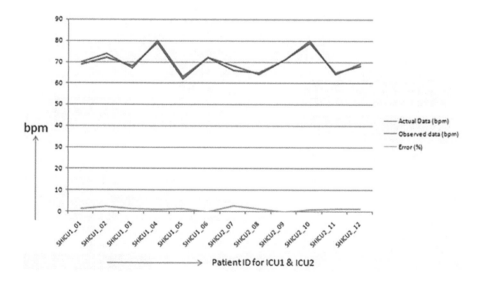

FIGURE 14.4   Performance analysis of system designed for bpm with error rate with existing system.

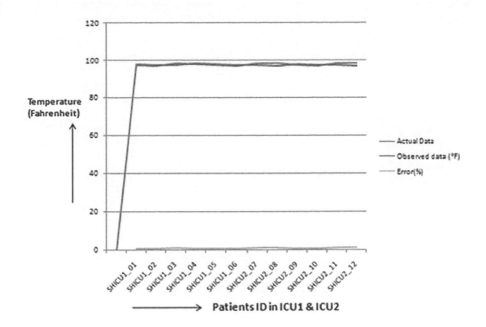

FIGURE 14.5 Performance analysis of system designed for body temperature and error rate with the existing system.

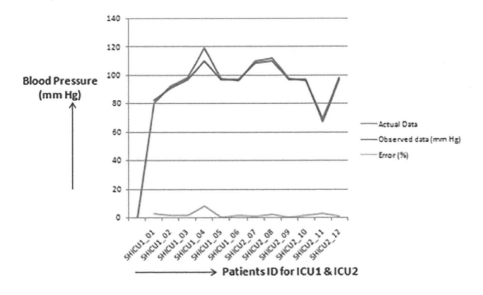

FIGURE 14.6 Performance analysis of system designed for blood pressure and error rate with the existing system.

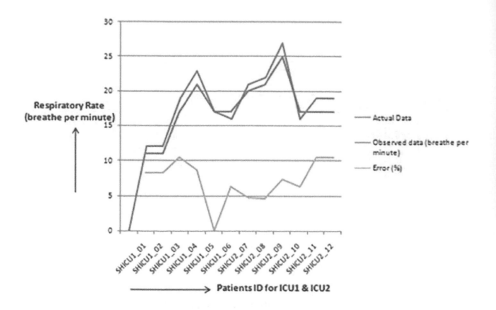

FIGURE 14.7 Performance analysis of system designed for respiratory rate and error rate with the existing system.

Based on the respiratory rate data collected by the actual machine available in ICU and IoT Smart Device designed, Figure 14.7 shows the performance analysis of the system designed for respiratory rate and error rate with the existing system. The divergence is due to motion artefacts generated by patient movement during treatment. Occasionally, the sensor is shifted, resulting in erroneous data. Furthermore, light dispersion from other sources produces divergence.

14.5 CONCLUSION AND FUTURE SCOPE

This research measures critical parameters like patient's heart rate, temperature, pulse rate and blood pressure actual and measured data using smart IoT-based innovative devices and ICU instruments in hospital. This device makes use of IoT, efficient wireless sensors and a high-precision microcontroller along with a wireless transmitter circuit that can be attached to the patient. The actual and measured data are well in agreement with each other. Further, this smart IoT device can be integrated with artificial intelligence and machine learning that will further increase the capabilities of this device in order to facilitate the healthcare and medical staff in real-time advanced technology.

REFERENCES

1. Jaiswal, S. P., Bhadoria, V. S., Agrawal, A., Ahuja, H. (2019). Internet of things (IoT) for smart agriculture and farming in developing nations. *International Journal of Scientific & Technology Research*, 8(12), 1049–1056.

2. Gandhi, P., Bhatia, S., Kumar, A., Alojail, M., Singh Rathore, P. (eds.) *Internet of Things in Business Transformation: Developing an Engineering and Business Strategy for Industry 5.0* (pp. 91–126). © 2020 Scrivener Publishing LLC.

3. Sahana, S., K. Singh. (2019). Fuzzy based energy efficient underwater routing protocol. *Journal of Discrete Mathematical Sciences and Cryptography*, 22(8), 1501–1515.

4. Aleksandar K., Natasa K., Saso K. (2016). E-health monitoring system. In Pece Mitrevski (ed) *International Conference on Applied Internet and Information Technologies*. Bitola, Macedonia.

5. Vishwakarma, S. K., Upadhyaya, P., Kumari, B., & Mishra, A. K. (2019). Smart energy efficient home automation system using IoT. In *2019 4th International Conference on Internet of Things: Smart Innovation and Usages (IoT-SIU)* (pp. 1–4). IEEE.

6. Islam, S. M. R., Kwak, D., Kabir, M. H., Hossain, M., Kwak, K.-S. (2015). The Internet of Things for health care: A comprehensive survey. *IEEE Access*, 3, 678–708.

7. S. Tyagi, A. Agarwal, P. Maheshwari. (2016). A conceptual framework for IoT-based healthcare system using cloud computing. In Abhishek Singhal (ed) *6th International Conference – Cloud System and Big Data Engineering* (pp. 503–507), IEEE, Location: Noida, India.

8. U. Dhanaliya, A. Devani. (2016). Implementation of e-health care system using web services and cloud computing. In Brijesh Iyer, S.L. Nalbalwar, R.S. Pawade (eds.) *International Conference on Communication and Signal Processing – (ICCSP)* (pp. 1034–1036). Atlantis Press: Lonere, Raigad (MS), India.

9. S. Biswas, S. Misra. (2015). Designing of a prototype of e-health monitoring system. *IEEE International Conference on Re-search in Computational Intelligence and Communication Networks (ICRCICN)* (pp. 267–272), IEEE: Kolkata, India.

10. A. M. Ghosh, D. Halder, S. K. Alamgir Hossain. (2016). Remote health monitoring system through IoT. *5th International Conference on Informatics, Electronics and Vision (ICIEV)* (pp. 921–926), IEEE: Dhaka, Bangladesh.

11. Brewster, C., Roussaki, I., Kalatzis, N., Doolin, K., Ellis, K. (2017). IoT in agriculture: Designing a Europe-wide large-scale pilot. *IEEE Communications Magazine*, 55(9), 26–33.

12. Yuehong, Y. I. N., Zeng, Y., Chen, X., Fan, Y. (2016). Internet of things in healthcare: An overview. *Journal of Industrial Information Integration*, 1, 3–13.

13. Nanayakkara, N., Halgamuge, M., Syed, A. (2019). Security and privacy of Internet of Medical Things (IoMT) based healthcare applications: A review. In *International Conference on Advances in Business Management and Information Technology*, Istanbul, Turkey.

14. Selvaraj, S., Sundaravaradhan, S. (2020). Challenges and opportunities in IoT healthcare systems: A systematic review. *SN Applied Sciences*, 2(1), 139.

15. Nazir, S., Ali, Y., Ullah, N., García-Magariño, I. (2019). Internet of Things for Healthcare using effects of mobile computing: A systematic literature review. *Wireless Communications and Mobile Computing*, 2019. https://doi.org/10.1155/2019/5931315.

16. Pulkkis, G., Karlsson, J., Westerlund, M., Tana, J. (2017). Secure and reliable Internet of Things systems for healthcare. In *2017 IEEE 5th International Conference on Future Internet of Things and Cloud (FiCloud)* (pp. 169–176). IEEE.

17. Premsankar, G., Francesco, M. D., Taleb, T. (2018). Edge computing for the Internet of Things: A case study. *IEEE Internet Things Journals*, 5(2), 1275–1284.

18. Kadhim T., Alsahlany, A. M., Wadi, S. M., Kadhum, H. T. (2020). Monitoring vital signs of human hear based on IOT. *Al-Furat Journal of Innovations in Electronics and Computer Engineering*, 1(2), 9–13.

19. Dash, S., Shakyawar, S. K., Sharma, M., Kaushik, S. (2019). Big data in healthcare: Management, analysis and future prospects. *Journal of Big Data*, 6(1), 54.

20. Sagar, R. H., T. Ashraf, A. Sharma, K. S. Raj Goud, S. Sahana, A. K. Sagar. (2021). Revolution of AI-enabled health care chat-bot system for patient assistance. In *Applications of Artificial Intelligence and Machine Learning*, (pp. 229–249). Springer, Singapore.
21. Al-Khafajiy, M., Kolivand, H., Baker, T., Tully, D., Waraich, A. (2019). Smart hospital emergency system. *Multimedia Tools and Applications*, 78(14), 20087–20111.
22. Mohammad, H. A., Chad, D. (2019). Design and implementation of an IoT-based smart home security system. *International Journal of Networked and Distributed Computing*, 7(2), 85–92.
23. School HM. *Tips to Measure Your Blood Pressure Correctly*. Harvard Health Publishing 2020. https://www.health.harvard.edu/heart-health/tips-to-measure-your-blood-pressure-correctly.
24. Kumar, A., Singh, V., Kumar, S., Jaiswal, S. P., Bhadoria, V. S. (2020). IoT enabled system to monitor and control greenhouse. *Materials Today: Proceedings*. https://doi.org/10.1016/j.matpr.2020.11.040.
25. Sambhi, S., Shikhar S., Bhadoria, V. S. (2021). IoT-based optimized and secured ecosystem for energy internet: The state-of-the-art. *Internet of Things in Business Transformation: Developing an Engineering and Business Strategy for Industry*, 5, 91–125. https://doi.org/10.1002/9781119711148.ch7.

15 Health Analysis by Digital Doctor Using Deep Neural Network

Amruta Khot
Walchand College of Engineering

CONTENTS

DOI: 10.1201/9781003248750-15

15.1 INTRODUCTION

Technology is improving its applications and the way we communicate with the rest of the world constantly. An efficient health-care system is important to a country's development, economy and industrialization. Health care is also an important determinant in ensuring the general physical and mental health and well-being of all the people around the globe. Medical records have to be stored appropriately. Also, all the prescriptions have to be stored properly. Thus, a developed application stores patients' past medical interactions in summary form. Also, it stores the prescriptions given by the doctor digitally.

Here, a developed video-calling web application enables digital interaction between the doctor and the patient using WebRTC (Web Real-Time Communications). WebRTC enables easy communication with transferring real-time audio, video and data in web. WebRTC is an open-source application programming interface (API) that enables peer-to-peer communication. Speech during the video call is converted into text in real time using Google Cloud Speech API. Google Cloud uses deep learning method to convert speech to text. Streaming speech-to-text API recognition is designed to capture the audio in real time and convert that audio to text, i.e. capturing and recognition of audio is done in real time [1].

Spoken language understanding refers to making machines understand human speech. It refers to targeted understanding for machines by converting inputs from the user into task-specific semantic representation. Slot filling is a technique in semantic parsing in spoken language understanding. It is a classification problem where every word from sentence has label. Recurrent neural network (RNN) is primarily built to implement slot-filling technique. Slot filling enables understanding the semantic context of words and filling in the slots to extract semantic concepts. Thus, for slot-filling technique, input is sequence of words that gives output as sequence labels of those words or slots filled for each corresponding word. Goal of our work is to find slot sequence for word sequence which has maximum posterior probability. The labels help to extract important keywords. Extracted keywords are used to generate summarized report for the patients. This report is stored in database thus enabling the patient to view report whenever required.

Recommendation system is used to display with similar symptoms of patients' health condition to doctor and also medicines prescribed to those patients. This enables the doctors to view what medicines were prescribed to patients earlier and thus doctor can prescribe appropriate medicines. Content-based recommendation system has been developed to recommend similar patients to doctor. Content-based systems hypothesize that if a patient has been prescribed a medicine in past, it may also be recommended to other similar patients in future. The designed module is content-based retrieval system that helps in recommendation that work on data given by the user explicitly or implicitly. Input to recommendation system is report generated of patient.

15.2 OVERVIEW AND PROBLEM STATEMENT

Academic study of natural language is a new discipline in science today. Doctors have to treat diseases as well as the patients, and thus language is an essential part of science. There are deficiencies in record keeping process of patients. Getting appointments and visiting doctor is tumultuous process. Critical information has to be stored properly. This information can further be required to perform diagnosis in future interactions during follow-ups. Thus, information about the previous interactions has to be stored safely. However, diagnosis information is stored by the patient in paper format. Thus, there is risk of losing the information due to various reasons.

Digital interactions can reduce the hassle of visiting the hospitals every time especially regular follow-ups. The patients need to meticulously maintain the prescriptions in paper format. The doctor should also have the details of the patient so that the doctor can review them whenever necessary. Many a time, patients may have to switch from one doctor to another. So, it becomes inconvenient for doctor as well as patient as doctor does not know the history of every patient. Video-calling web application will help patients and doctors to interact with each other. Speech extracted from the video conversations is converted into text in real time. Thus, conversations are converted into text format. Clinical information from the narrative medical interactions is extracted and saved for the patient. Thus, concepts will be extracted from the conversations. It would help to classify the relationships between the concepts (for example, symptoms and duration of the symptoms) and identify the nature of assertions about the concepts(for example, assertive questions such as does patient have headache and answer for this can be yes or no only). This keyword or information extraction will be stored for the patient so that it will always be available to him.

Numerous medications are available for a particular disease or symptom. However, not all medications work for all the patients. Thus, doctor will be recommended patients, similar to the current patient based on information extracted and also the medications or prescriptions provided to those patients. Doctor can then prescribe medicines to the patients accordingly by referring the similar patients. Also, these prescriptions will be available on patients' profile which will save the patients' efforts of saving the prescriptions in paper format.

15.2.1 MOTIVATION

Medical dialogues between doctor and patient play a crucial part in medical diagnosis. Studies have shown that 80% of diagnostic assessments are based on the medical interactions. Also, there have been tremendous advancements in medical domain.

Significant amount of time of patients is spent in visiting the hospitals for normal routine follow-up. Also, there may be a case when the doctor is not available. Thus, patients have to wait for the doctors to get back in town. Hence, to make significant use of technological advancements, a developed web application will enable the patients to reach out to the doctors easily and conveniently.

Unprecedented amount of information relating to patient's symptoms, diseases and treatments is produced through the interactions. However, not all information can be considered to be significant. Natural language processing (NLP) can help in processing that enormous information and extracting only the relevant information from the conversations. Thus, audio from the video interactions is converted to text in real time and processed.

Numerous medications are available for each disease. Thus, in order to prescribe a medication, doctor is displayed patients with similar medical condition so that doctor can know if a medicine is more effective on the current patient being examined. Recommendation of similar patients thus helps doctor in knowing whether particular medication had been actually effective for patients.

15.2.2 CONTRIBUTIONS OF THIS WORK

Capturing significant and accurate data from clinical interactions holds a key in maintaining the integrity of practicing medicine. The health-care enterprise worldwide produces massive amounts of records relating to patient situations, sicknesses, signs, symptoms and remedies in just a day than what ought to probably be understood in a lifetime. Keywords extracted from the interactions can be used to store patient-related data. However, doctors or even the patients don't want to review the conversations again in case they want some important information from the conversations.

Most video-calling applications have to be either installed by the users in order to communicate in real time or a plugin has to be installed for a particular browser. These plugins are also browser dependent. Hence, when user is using different browser, new plugin has to be installed. Thus, developed video-calling web application using WebRTC will enable the doctors and patients to easily communicate over web. It enables exchange of voice and video data transmission between the doctor and the patient. WebRTC is browser-independent. It comes with a JavaScript API layer on the top so that it can be used in the website. Channels are created using WebRTC API which enable flow of information. Creating a unique room, accessing the network information and streaming data have to be performed using WebRTC APIs.

Speech-to-text conversion had been introduced with the idea that users can have human-like interaction with the system. Google Cloud Speech-To-Text JavaScript API is available for commercial use and can be used to convert the audio or speech in the video-calling interaction into text in real time. Also, it is freely available to be used by the developers. Thus, WebRTC API had to be combined with Google Cloud Speech-To-Text API so that the video interactions can be converted into text.

Natural language understanding is used in extracting important utterances in the text or conversations. A main aim in spoken language processing in purpose-orientated human-machine conversational knowledge structures is to interpret semantic ideas, or to fill "slots" inserted in a semantic meaning, to be able to gain a goal of understanding context in a human-machine speak. Let us consider the examples in Table 15.1.

Thus, labels have been added corresponding to each word. Also, model does not need the patient to answer in complete sentences, i.e., it can understand the context from the rhetorical questions as well. Word embedding converts words into continuous vector representation by mapping semantically similar words close to each other in the space and dissimilar words comparatively far from each other.

RNNs are designed to use the previous information, i.e. in neural networks, consider that inputs are not related to each other. Slots are to be filled depending upon the previous input statement and the slots filled for those words. Long sequences

TABLE 15.1
Slot Filling

Slot Filling I

Words	I	have	been	having	head-ache	since	last	three	days
Labels	0	0	0	0	A-symp	0	0	A-symp-dura	B-symp-dura

Slot Filling II

Words	Do	you	have	cough	no
Labels	0	0	0	A-not-symp	A-neg

Slot Filling III

Words	Are	you	Suffering	From	any	allergy	yes	dust	Allergy
Labels	0	0	0	0	0	A-All	A-posl	A-All	A-All

can suffer from vanishing gradient problems which can be avoided by using gated recurrent units. Gated recurrent units decide as to what amount of information has to be passed from previous computations. Also, convolutional lookahead layer has been used to capture information in current timestamp by using only some amount of future information. Besides, adding convolutional lookahead layer instead of bidirectional RNN layer is that it reduces the computations being done.

Model has been fed two types of input and thus same model has been analyzed for different types of input. The first approach uses words that are present only in the dataset, whereas the second approach uses all the words in the dictionary.

Patients with similar medical conditions are displayed to doctor. Also, the medications given to those similar patients are displayed. Doctor can refer medications given to similar patients and prescribe medications. Similar patients are searched using content-based recommender system. This system predicts the content or next sentence in sequence based on context through the user audio or video data.

15.3   TERMINOLOGY

Humans have to store their medical reports in paper format. Besides, sometimes doctor may be unavailable in the town. However, they can connect virtually. Natural language understanding is focused on enabling the machines to understand and interpret the human language. Goal is to convert user input into task-specific semantic representation of users' intentions. Slot-filling technique is used as a sequence classification problem to assign labels to each of the words, i.e., performing tagging for each word.

Natural Language Processing: NLP helps the computer to manipulate and interpret human language [2,3]. It fills the gap between the computer and human communication. It allows the computer to perform different tasks based on inputs given by the user in human language. It is being used almost in every aspect of human life today. Few areas in which NLP is being used on large scale are:

- Intelligent robots
- Checking for spam email messages
- Speech recognition
- Language translation
- Keyword extraction

15.3.1   NATURAL LANGUAGE UNDERSTANDING

Natural language understanding is aimed at extracting the meaning of speech utterances. It is used to understand conceptual meaning of natural language sentences. The term has been designed for understanding the context of human speech to machines. Natural language understanding is a component of NLP. Natural language understanding allows the computer to statically understand human languages without the use of if/else. Complexity of NLP systems is measured on two metrics, breadth of the system and the depth of the system [4]. Breadth of the system signifies the vocabulary size whereas the depth of the system signifies how accurately system understands. It goes beyond NLP by its ability to understand incomplete sentences, rhetorical question-answer system, etc.

Keyword Extraction: Keywords are the shortest units of content that can be used to summarize the document. Keyword extraction technique is used to extract important information from text. Keyword represents concise information about the data. Basic idea of keyword extraction system is to extract the handful of keywords from continuous bag of words, thus representing small word set from full list of words.

Recurrent Neural Network: RNNs are a variant of neural networks that take input from previous outputs. Traditional neural networks don't have connection between the inputs. However, sometimes there is connection between the inputs. RNNs contain a hidden layer which stores the information about the previous sequence in order to predict the output of the current input. RNNs transform independent activations to dependent activations and thus give same weights and biases input to all the layers. Thus, current output depends upon the previous computations. It contains "memory" to capture information captured so far, i.e. capture slots filled so far. RNNs remember each and every information at every timestamp.

Gated Recurrent Unit: RNNs suffer from vanishing gradient problem. Gated Recurrent Units aims to help vanishing gradient problem. GRUs decide what information to be passed ahead instead of passing all the information. Thus, they are trained to keep information from long time with washing away important irrelevant information from past.

Slot Filling Technique: A major task in NLP is to define semantic concepts by filling in slots or set of arguments in order to achieve its goal. Thus, it is sequence classification problem in which each word is tagged with a slot or label. Thus, the input is collection of word in the form of sentence and generates output as sequence of labels for those words.

Content-Based Recommender System: This system is used to predict user's performance rating for a particular item [5]. Content-based recommender system predicts an item based on previous similar item. Thus, basic idea is if a person likes a particular item, person will also like similar item. Thus, it makes predictions based on past user predictions. Content-based recommender system relies on similarity of the items.

15.4   OBJECTIVES

15.4.1   To Develop a Video-Calling Web Application

Friendly interface is required to understand the need of users and interpret their perspectives and build an environment where they can easily make use of the available features. The novice need not learn any tool or technique to use the application of this interface in their daily life routine. The attributes to be used would be very common and friendly enough for anyone to handle.

15.4.2   To Develop a Video-Calling Web Application

This module aims to convert the information collected from the video call into a required format so that the information can be further processed and made understandable by the machine. It should be able to extract the audio from the video call conversation and then convert that audio into simple text in real time. This simple text is to be then converted into dialogues that are to be grouped with each other in such way that it could be provided as an input in vector form for the neural network model to understand.

15.4.3 To Implement Feature Extraction so as to Extract the Patient's Details to Develop a Video-Calling Web Application

It is very essential to understand the conversations between the doctors and the patients if the aim is to finally generate a medical report based just on the conversations. The objective is to create a module that will understand the dialogues generated and then create a report extracting necessary information from these dialogues.

For this purpose, it needs to understand various features of NLP that would help it to extract just the relevant information and discard the other. It needs to learn to identify which of the words are of importance and what does that particular word depicts. This would help it to provide meaning to the sentences in the dialogue. Once it figures out the meanings of the sentences, now the final task is to generate the report taking into consideration information that has been extracted yet.

15.4.4 To Build a Recommender System Similar for Patients Based on the Features Extracted

The ultimate objective is to display the profiles of the patients that are similar to the one that the doctor has been currently working on. The aim is basically to create a content-based recommender system to provide recommendations to the doctor.

First of all, the model must understand the features of the current patient. Then, it must be able to compare those features with the history of all other previous patients and then display the cases that have the most similarity score.

15.5 PROPOSED METHOD/ALGORITHM

This work aims in making a doctor-patient conversation to be easily and conveniently carried out. The idea is to create a video-calling web application between the doctor and the patient so that they can communicate with each other from where ever they are and at a time that is feasible to both. It also helps to generate a medical report automatically understanding the conversation between them and retaining required documents whenever necessary.

Consider a patient is having a medical emergency and he/she has to visit the doctor but the doctor is out of town. This is one of the cases where the patient can contact the doctor through video call and the report of their conversation will be automatically generated. Slot filling, a technique to understand spoken language and particularly the dialogue, has been used for the generation of the report by understanding and extracting important keywords from the conversations. The term spoken language processing for semantic analysis is understanding of human speech directed at machines. The process to implement for transformation of user input which is sequence of words to semantic representation giving meaning or intention of user input as dialogue manager takes the as context of user input and try to interpret for taking action that gives semantic meaning and context of user input with meta information like geo-location, any personal preference.

Let us have an example. Consider the following conversation:

Doctor: Do you have
Patient: Yes

This is a basic conversation between a doctor and a patient. There are some keywords in this conversation that would help the machine to understand necessary information to generate the report. The keywords are *"fever"*, a symptom and *"Yes"*, a positive response. Through these keywords, it would understand that the concerned patient is having a fever and accordingly it would be displayed in the report. Similarly, all other medical parameters would be considered to extract significant information and generate a report. The doctor will be then provided with recommendations of similar patients that would be helpful in providing instructions to the patients.

15.5.1 PROBLEM DEFINITION

The designed system is proposed to reduce the communication gap between a doctor and a patient. Traditionally, a patient has to physically visit a doctor in case of any medical situation, but it may not be feasible either to doctor or the patient to do so every time. Mutual time management is also one of the problems in this case. Generation of reports, maintenance of some important documents, and referring previous similar cases are somewhat tedious and time-consuming job for a doctor.

To overcome this problem, the following system is proposed, which will try to create a new system making things better.

15.5.2 PROPOSED IDEA/SYSTEM

Generation of a video-calling web application between a patient and a doctor and then generating a report automatically through their conversation: This system is divided into several parts. The first part is a video-calling interface in which the conversation will be converted to text from extracted audio in real time using speech-to-text conversion. The audio is converted into text in real time in a format required for further operations. Slot filling is used to understand the text and then generate a report automatically. The report is saved in the patient's profile and the doctor is provided with recommendations of similar patients to make it easy to give instructions (Figure 15.1).

The proposed system has the following features:
Video calling between the doctor and the patient.

- Converting audio to text in real time.
- Understanding the text and extracting significant features from it using slot-filling algorithm.

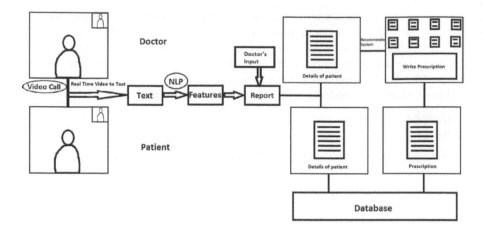

FIGURE 15.1 General flow chart.

- Generating a report from extracted features and doctor's inputs whenever essential.
- Store the report to doctor's as well as patient's profile, so that any one of them can access it digitally anytime.

Generation of an interface for video calling between the doctor and the patient: WebRTC is to capture web applications and sites to capture and buffer audio and/or video media, and to exchange data between browsers without requiring an intermediary [6]. The audio from this video call is then converted to text in real time. We have used Google Cloud Speech-to-Text to convert audio to text [7,8].

The RNN model is used for classifying word as "*keyword*" or "*not*". Feature values have been used for classifications. A major task in NLP is to extract semantic concepts of words and sentence and then to fill set of arguments or\slots embedded in a semantic frame. For the slot-filling task, the input is the sentence consisting of a sequence of words, and the output is a sequence of slot/concept IDs, one for each word. Slot filling algorithm has been used for this part.

$$r = \frac{\sum_{i=1}^{n}(x_i - X)(y_i - Y)}{\sqrt{\sum_{i=1}^{n}(x_i - X)^2}\sqrt{\sum_{i=1}^{n}(y_i - Y)^2}} \tag{15.1}$$

The generated report is then stored in the patient's profile for further use and then using the features extracted from the conversation, the doctor is recommended with cases similar to the current one. For this part, a content-based recommender system is created.

15.6 SYSTEM ARCHITECTURE

Figure 15.2 clearly displays the architecture used to model the system. Beginning with the modeling of the friendly user interface and moving toward the back end of

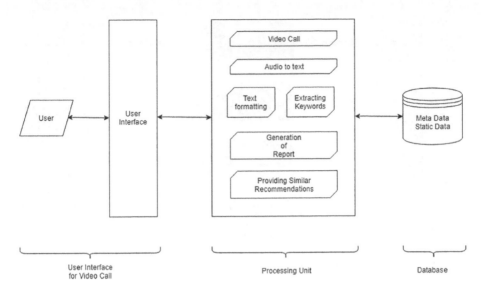

FIGURE 15.2   System architecture.

the layout to generate the medical report, the processing unit explains it all on the procedures involved from video call to report generation. Then recommendations of the similar patients also include connectivity with the database to extract the content.

15.6.1   System Analysis and Design

1. **User Interface:** A web application that allows the users to connect with one another through a video call
2. **Video Call:** A medium of conversation between the doctor and the patient
3. **Audio to Text Conversion:** The audio is extracted from the video call and is converted to text in real time
4. **Text Formatting:** The text retrieved from the video call is then converted in a set of dialogues in a suitable format that can be provided as an input to the neural network for understanding
5. **Extracting Keywords:** Essential keywords are extracted from the formatted text using slot-filling algorithm of Natural Language Understanding implemented using neural networks
6. **Generation of Report:** Finally, using these extracted keywords, a medical report is been generated that contains all the necessary information
7. **Providing Similar Recommendations:** The doctor is immediately recommended with cases of similar patients based on the report generated
8. **Data Storage:** The report is then stored in the database to be used in the future

15.6.2   Requirement Specification

The interface is compatible with all platforms that support web application and thus making it highly portable as well as compatible. It is reliable enough for any

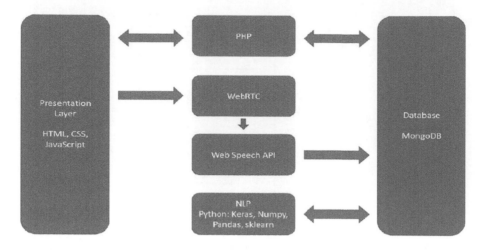

FIGURE 15.3   Technologies used.

organization to integrate with their database and achieve the optimal results without any ambiguity in the system and its data stored. Expected errors and anomalies have been considered and handled in case of exceptions which make it robust enough for any layman to use it without a second thought.

Hardware support is not much of a cause of concern as any system able to carry out a video call can launch this system and there are no such system configurations required as the data size has been handled using the server and the compatibility of the interface would not be an issue. Scalability and reliability are the most important features which make this implementation an effective outcome.

Technological Requirements (Figure 15.3):

System Requirements:

1. Web camera
2. Microphone
3. Browser

Use Case Diagram (Figure 15.4):

15.6.3   DESIGN AND TEST STEPS

1. Connecting video call between doctor and patient and carrying out the conversation
2. Extracting the audio from the video call and converting it into text in real time
3. Applying NLP on that extracted text to extract essential keywords for understanding the conversation
4. Generating and displaying a medical report using the extracted keywords
5. Storing the report in the database for future
6. Implementation of a content-based recommender system to recommend the doctor with similar patient's cases

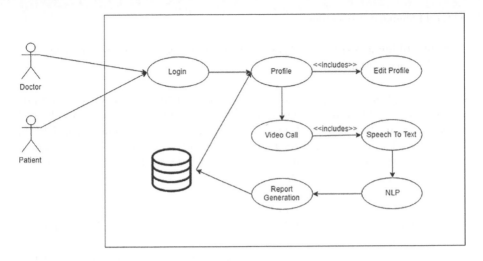

FIGURE 15.4 Use case diagram.

15.6.4 ALGORITHMS AND PSEUDO CODE

This work provides a platform for video calling between a doctor and a patient and automatically generates a medical report understanding the conversation between the doctor and the patient. It also recommends the doctors with cases of similar patients after the conversation so that the doctor may refer. The procedures followed in the design and development respective model are simple and accurate making it an efficient step-by-step procedure for the code and its implementation to be understood.

Video Calling

- WebRTC API used to implement video calling
- A virtual room will be created where the doctor and the patient will join for the video call
- Interactive Connectivity Establishment (ICE) is used to allow WebRTC to overcome the complexities of real-world networking
- Session Traversal Utilities for NAT (STUN) and Traversal Using Relay NAT (TURN) servers have been used
- In case the STUN fails, ICE uses TURN

Speech to Text

- Speech from conversation from the video call is converted to text using Web Speech API
- This conversion takes place in real time
- The main interface for Web Speech API is speech recognition
- The doctor and patient dialogues will be then converted into required format considering timestamps

Natural Language Processing

- Slot-filling algorithm has been used to extract essential information from the conversations

Slot-filling algorithm is supervised learning model which gives each word appropriate labels as to what information the word holds as this is supervised which need some dataset to train the model upon. For this, the dataset is created and labeled accordingly. The machine understands only numbers and not the text. Therefore, before feeding this dataset to the machine/model to learn, it must be converted to vectors of numbers.

- It is a bidirectional RNN implemented using neural networks
- All the words in the word-dictionary are been considered while designing the model
- The text extracted from the speech is then provided as input to the model and the output, i.e. the labels along with the original words are stored in a file
- This file is then passed through another set of code that generates a medical report depending upon the labels generated by the neural network model
- This report includes all the essential details and information extracted from the conversation. For example, the report may include the symptoms that the patient may be suffering from, or any suggestions or prescription that the doctor may provide, etc.
- The report is displayed to both the doctor and the patient and also stored in their profiles.

Recommender System

- A content-based recommender system has been used to recommend similar cases of patients to the doctor after the conversation is done
- Keywords extracted from the slot-filling algorithm have been used to build the recommender system
- The keywords extracted for the current patient's case are compared with the keywords of previous patients
- The result is the similarity score of all the patients with respect to the current patient
- Top 10 patients with the highest similarity will be displayed to the doctor

15.7 PERFORMANCE STUDY

15.7.1 IMPLEMENTATION

Implementation of the work is divided in four parts viz. video calling, speech to text, NLP and recommender system.

15.7.2   VIDEO CALLING

To implement video calling, WebRTC APIs are used. While starting the video call, the patient clicks on *"Call a doctor"* button. At this time, a random hash code is generated which is used to recognize the room created for doctor and patient. When doctor opens the link containing the hash code, doctor will join to the room created. Using ScaleDrone, a channel can be subscribed directly. Here, the channel is used with ID "yiS12Ts5RdNhebyM".

ICE allows WebRTC to overcome the complexities of real-world networking by finding the best path to connect two peers either by direct connection between the clients or when it's not possible to have direct connection. In case of NAT, ICE uses STUN server. This allows clients to _nd their IP and type of NAT in use. In case the STUN fails, ICE uses TURN. Unlike STUN, TURN remains in the path found by ICE even after connection is made. This makes the network slower and not having to use TURN is always desirable (Figure 15.5).

For the system to find the STUN and TURN servers, RTCPeerConnection is used with valid configurations.

The configuration looks like:

```
Const configuration = iceServers: [ urls: 'stun:stun.l.google.
com:19302'];
```

ICE finds the URLs mentioned for STUN and TURN. After connection, the events *"onicecandidate"*, *"on-addstream"* and *"onnegotiationneeded"* are handled which enable actual video calling. To access the user media, i.e. camera and microphone, we use method named *getUSerMedia()*. This asks the user for permission for accessing the camera and microphone. The system works if and only if the user gives the consent.

15.7.3   SPEECH TO TEXT

Speech from conversation from the video call is converted to text using Web Speech API. Since, this happens real time, the module uses the speech from the microphone

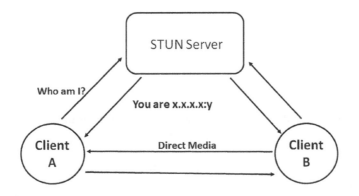

FIGURE 15.5   STUN and TURN servers.

directly which saves the time to record and save the speech. Main interface for the Web Speech API is "SpeechRecognition".

After instantiating an object of "SpeechRecognition", user permission for microphone is needed. This is not necessary because already have the required permission from the video call module. The method *start()* of the "SpeechRecognition" starts the speech recognition process. When a valid word is recognized, the method will trigger event which contains the result. This event is handled by method *"onresult()"*. The patient's and doctor's dialogues will be converted into an appropriate format considering timestamps that would be suitable for the processing.

15.7.4 NLP

For extracting the features from the text generated from speech-to-text module, slot-filling algorithm is used. Slot-filling algorithm is supervised learning model which gives each word appropriate labels as to what information the word holds. As this is supervised, we need some dataset to train the model upon. For this, the dataset is created and labeled accordingly, e.g.,

BOS *i* have upset stomach EOS 0 0 0 *A-symp B-symp* 0

Above is one of the sentences in the dataset. BOS and EOS denote beginning and end of the sentence, respectively. Here, in the sentence, *"upset stomach"* is the symptom which the patient is telling while *"i"* and "have" do not contribute in recognizing the symptom. Thus, *"i"* and "have" are labeled zero. Also note that upset stomach is only one symptom and not two different symptoms even if they are two different words. Therefore, labels $'A-symp'$ and $'B-symp'$ are used. $A-symp$ denotes start of the label *"symptom"* and $B-symp$ denotes up to where that symptom is. The dataset contains 670 unique sentences and 876 unique words. The machine understands only numbers and not the text. Therefore, before feeding this dataset to the machine/model to learn, it must be converted to vectors of numbers. Each word is given a unique index. Then, the word from the sentence is replaced by this unique index. Thus, a sentence is converted to a vector of numbers where each number corresponds to a unique word. Like words in the sentence, labels are also converted to the indexes and output is in form of vector of indexes of labels.

This is implemented in two ways. In the first approach, only unique words from the dataset are assigned unique index. This keeps the size and range of indexes in limit but when new word is discovered outside of the dataset, the model does not find index for the word. In the second approach, all words in English dictionary are assigned an index. This increases the range of the dictionary drastically and therefore time required to train the model.

Architecture of the model used to train the machine to identify the labels is given in Figure 15.6.

Apart from input and output layers, the neural network model has four hidden layers.

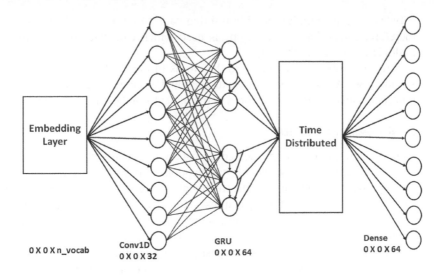

FIGURE 15.6 Architecture of the model.

Embedding Layer: The indexes are assigned completely random. Semantic relation between the words is not portrayed by the index assigned to them. For this problem, first the embedding layer is used which according to the vector of each word assigns new vector with updated indexes. This makes similar words have closer vectors. In short, this layer groups the words according to the meaning of the word.

Convolution Layer: Bidirectional RNNs are challenging to deploy in online low latency settings. To counter this, a lookahead convolution layer is used which gets information from future subsequence in efficient manner to improve unidirectional RNNs.

Gated Recurrent Unit: GRU is an extension of Long Short-Term Memory (LSTM) which is an extension of a simple RNN.

GRU works on two gates, update gate and reset gate. In update gate, which information is discarded and which information is kept is decided while reset gate decides point in past up to which information should be kept. Since the created dataset is limited, the use of LSTM is not advisable which needs larger dataset. For this reason, a layer of gated recurrent units is used. Also, due to lesser tensor operations, GRUs are faster to train.

Time Distributed Dense Layer: This layer is used for each word to determine the appropriate label for the word. Time distributed dense layer is used because it needs to apply fully connected dense on each time step and get output separately by time steps which helps us to keep one-to-one relations on input and output. The softmax function calculates probability of each label for the word and label with the highest probability is chosen. The type of gradient descent used for the model is Adam and loss function used is categorical cross-entropy. Adam is an optimization algorithm that is extension of the classical stochastic gradient descent procedure to update network weights iterative based in training data. More than two labels and therefore, categorical cross-entropy is used which calculates the loss for multiple categories of the output.

After the labels are assigned to the words through NLP model, corresponding words of each label are extracted and a report is generated. This report is displayed as well as stored to the patient's and doctor's profile.

15.7.5 RECOMMENDER SYSTEM

The recommender system is implemented to display the details of similar patients to the doctor while the doctor writes the prescription. This will enable the doctor to review other patient's symptoms and the treatment they received. The content-based recommender system is used. This system uses metadata like other patient's symptoms, calculates similarity to the current patient and based on the similarity score, decides whether to recommend the patient to the doctor or not.

For the implementation, some data of past patients should be available. Like in NLP, the symptoms are converted to a vector and this vector is input to the recommender system. The result is similarity score of all patients with reference of current patient. The highest ten patients will be displayed.

First, TF-IDF (Term Frequency-Inverse Document Frequency) vector of each patient's, current as well as past, symptom from report is calculated. This vector is used to calculate similarity score. To calculate this, method linear kernel() is used.

15.8   RESULTS AND ANALYSIS

This section contains the screenshots of the results of each module.

In Figure 15.7, the output of speech-to-text module which is also stored for further process.

FIGURE 15.7   Result of speech-to-text module.

A peak of dataset is shown in Figure 15.8. The dataset contains the sentence along with the labels.

Figures 15.9 and 15.10 show the $F1$ score of implementation of two approaches to input data.

From Figures 15.9 and 15.10, it is observed that $F1$ score of the first approach is lesser than that of the second approach.

Figures 15.11 and 15.12 show the process from input file to report generated after assigning the labels.

In the above figures, the labels are assigned to each word by the model and stored in a file. These labels will be used to generate report.

FIGURE 15.8 Dataset.

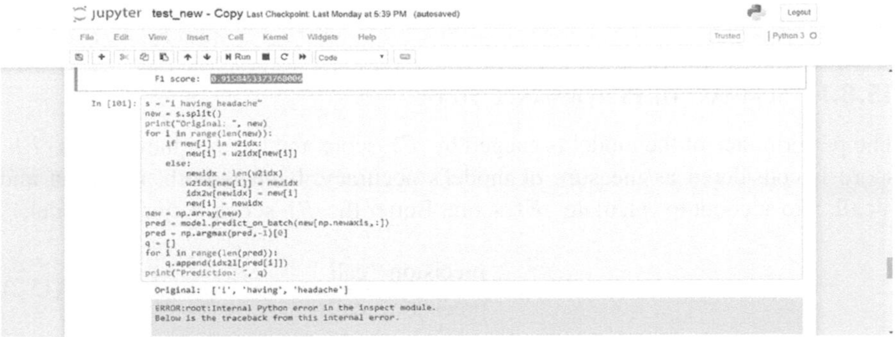

FIGURE 15.9 Result of the first approach with words from dataset.

```
[83] #print(len(val_pred_label))
     f1 = precision_recall_fscore_support(val_label_list,val_pred_label, average = 'micro')
     print("F1 score: ", f1[2])

 ⊏→  F1 score:  0.9301075268817204                                                          ⌄

[84] model.save_weights('/content/gdrive/My Drive/Colab Notebooks/DataFinal 100% accuracy1 Copy.h5', overwrite = 'true')

 ▶   s = "i do not have fever from nine days"                                                ⋮
     new = s.split()
     print("Original: ", new)
     for i in range(len(new)):
         if new[i] in w2idx:
             new[i] = w2idx[new[i]]
         else:
             newidx = len(w2idx)
             w2idx[new[i]] = newidx
             idx2w[newidx] = new[i]
             new[i] = newidx
     new = np.array(new)
     pred = model.predict_on_batch(new[np.newaxis,:])
     pred = np.argmax(pred,-1)[0]
     q = []
     for i in range(len(pred)):
         q.append(idx2l[pred[i]])
     print("Prediction: ", q)

 ⊏→  Original:  ['i', 'do', 'not', 'have', 'fever', 'from', 'nine', 'days']
     Prediction:  ['0', '0', '0', '0', 'A-not_symp', '0', 'A-not_symp_dura', 'B-not_symp_dura']
```

FIGURE 15.10   Result of the second approach with words from the English dictionary.

FIGURE 15.11   Sample input data.

15.8.1   SUMMARY OF PERFORMANCE STUDY

The performance of the model is gauged by $F1$ score and mean squared error. $F1$ score is considered as measure of model's accuracy. It takes both, precision and recall, into account to calculate $F1$ score. Better the $F1$ score, better the model.

$$F1 = 2 * \frac{\text{precision} * \text{call}}{\text{precision} + \text{call}} \qquad (15.2)$$

```
1   please come in thank you      0 0 0 0 0
2   what have you come in for today i have been having some pain in my joints especially the knees  0 0 0 0 0 0 0 0 0 0 0 A-symp 0 0 B-symp 0 0 0
3   how long have you been having the pain i would say it started three or four months ago it is been getting worse recently    0 0 0 0 0 0 0 0 0 0 0 0 0 A-symp_dura
4   are you having any other problems like weakness fatigue or headaches well i have certainly felt under the weather   0 0 0 0 0 0 0 A-symp A-symp 0 A-symp 0 0 0 0
5   right how much physical activity do you get do you play any sports some i like to play tennis about once a week i take my dog on a walk every mornin    0 0 0 A-p
6
```

FIGURE 15.12   Labels assigned.

TABLE 15.2
Results

	First Approach	Second Approach
F1 score	0.9158453373768006	0.9301075268817204
Mean squared error	32.83717357910906	28.821812596006144

Mean squared error is average of the squares of the error. These values are always positive. Mean squared error values closer to zero are desired.

From Table 15.2, the first approach in which used whole English dictionary with a better F1 score and lesser mean squared error. This is due to the first layer used in the model that is embedding layer. More words made the embedding more effective and the vectors of each word are more aligned with similar words.

15.9   CONCLUSIONS AND FUTURE WORK

NLP is a powerful tool which will make revolutionary changes in the field of Artificial Intelligence. With the help of natural language understanding, information extraction from unstructured languages like English becomes easy and will help build more tools to make humans more compatible to computer system. The tool is build using power of NLP and recommender system making medical processes easy. It can not only understand doctor-patient conversations but also work on any other datasets.

It is built to understand simple conversations between doctor and patient and automatically create medical report based on that understanding. It also recommends doctor with profiles of patients similar to the current case. In future, the system can be improved by taking complex conversations into considerations and also understanding the problem by itself and providing the solution without consultation of the doctor. With the continuous evolution of NLP, more techniques will rise and development of such system will increase.

REFERENCES

[1] V. Kumar, M. L. Gavrilova, C. J. Kenneth Tan, and P. L'Ecuyer, "Preface," *Lect. Notes Comput. Sci. (including Subser. Lect. Notes Artif. Intell. Lect. Notes Bioinformatics)*, vol. 2667, no. August, 2003, doi: 10.1007/3-540-44839-X.

[2] X. Yang and J. Liu, "Using word confusion networks for slot filling in spoken language understanding," *Proc. Annu. Conf. Int. Speech Commun. Assoc. INTERSPEECH*, vol. 2015-January, no. 3, pp. 1353–1357, 2015, doi: 10.21437/interspeech.2015-47.

[3] M. Mahdavi, "Implementation of a recommender system on medicalrecognition and treatment," *Int. J. e-Educ. e-Bus. e-Manag. e-Learn.*, vol. 2, no. 4, pp. 315–318, 2012, doi: 10.7763/ijeeee.2012.v2.136.

[4] "WebRTCAPI–WebAPIs–MDN."https://developer.mozilla.org/en-0AUS/docs/Web/API/WebRTCAPI.

[5] "Real time communication with WebRTC." https://codelabs.developers.google.com/codelabs/webrtcweb/%0A0.

[6] "Cloud Speech-to-text-Google Cloud." https://cloud.google.com/speech-to-text/.

[7] Negi, K., G. P. Kumar, G. Raj, S. Sahana, and V. Jain. "Degree of accuracy in credit card fraud detection using local outlier factor and isolation forest algorithm." In *2022 12th International Conference on Cloud Computing, Data Science & Engineering (Confluence)*, pp. 240–245. IEEE, 2022.

[8] Anand, A., S. P. Mishra, and S. Sahana. "Assistive devices and IoT in healthcare functions." *Deep Learning and IoT in Healthcare Systems: Paradigms and Applications*, p. 103, 2021.

16 AI and IoT in Supply Chain Management and Disaster Management

Kaljot Sharma and Darpan Anand
Chandigarh University

CONTENTS

DOI: 10.1201/9781003248750-16

16.1 INTRODUCTION

The world has progressed toward a new transition over the years, and Industry 4.0 technologies are considered as the way of the future. Artificial intelligence (AI) is one of the most well known of these technologies, which is defined as a computer's capacity to interact with and imitate human abilities (including block-chain, Internet of Things (IoT), cloud computing, and others). When AI is employed, issues are addressed with greater precision, speed, and number of entries [1]. AI is neither a new concept nor an academic area. AI is emerging as a competitive advantage, whereas certain aspects of information technology are being reduced to a competitive necessity. Many firms are shifting from remote monitoring to control, efficiency, and, eventually, sophisticated autonomous AI-based technologies in order to improve their performance. It also has a lot of clout in the corporate world [2]. AI has a growing and broader presence in academic literature, and its presence has influenced several more sectors, including research project, that have selected up on the facts, and AI has been studied more from a wider view, with supply chain management (SCM) identified as one of the areas most likely to benefit from Application domains. Despite the fact that research and practice are enthusiastic about AI (as indicated by the lot of researches on AI), there is a need to study AI's contributions to the field of SCM [3]. What role does AI play in SCM research. The management of a pattern of interactions within a company as well as between interrelated firms and business modules comprising third-party vendors, acquiring, manufacturing plants, transportation, advertising, and linked facilitates the forward and invert supply of resources, facilities, funding, and data from the original creator to the end user with advantages of providing value chain with the use of IoT: RFID, Scanners, and other IT-linked sensor products that are distinguishing traditional supply chain with the use of IoT: RFID, Scanners, and other IT-linked sensor products that are distinguishing traditional supply chain with smart supply chain and attempting to convert the entire procedure easy, comfortable, and pliable [4,5].

The IoT revolution is transforming every organization and the logistics sector is no exception [6]. IoT can efficiently give certain different resources to diverse domains throughout the supply chain journey. The IoT movement is reshaping every business, leaving no stone unturned, and the logistics industry is no different. Throughout the supply chain journey, intelligence in certain regions, new information may be obtained. By adding intelligence in certain regions, new information may be obtained. IoT allows logistics companies to operate more effectively and provide clients with more tailored, intelligent, and automated solutions via a system that does not require human assistance or involvement [7].

16.1.1 SUPPLY CHAIN MANAGEMENT

Traditional SCM, which depended on manual and tedious data gathering and commodity tracking, has developed through time. A business creates a network of distributors (or "links" in the chain) to transport the commodity from suppliers and manufacturers to organizations that interact directly with the customer [8,9].

Distribution network, logistic support, stock management, and storage are all phrases that are used interchangeably in the industry. All enterprises in the supply

FIGURE 16.1   Supply chain activities can be served through IoT.

chain are connected by technology and logical processes. Physical flows include the conversion, distribution, and storage of products and resources. These are by far the most visible and recognizable supply chain components. Various companies and departments along the supply chain use information shared to coordinate their long-term goals and govern the daily movement of goods and commodities from supply chain providers [10]. All the activities are mentioned in Figure 16.1.

16.2   WHAT IS IOT?

The IoT is a network of organized networked computer smart devices, products, automobiles, machines, and people linked by Unique Identifiers [11]. To communicate data via the public network, no human transmission interaction is needed. A network with a wearable device, a farm mammal with a bio processor transceiver, a vehicle with built-in sensing devices to alert the driver when a wheel level is low, or any natural or man-made entity. Organizations in a variety of industries are progressively using IoT to function more efficiently, better understand customers in order to deliver improved customer care and response capabilities, and boost brand profitability [12,13]. Security for these applications is also important [14–16].

16.2.1   ROLE OF IOT IN SCM

Major goals of IoT implementation in SCM are tracking and monitoring. However, the IoT is capable of much more than wealth management. The following are some significant insights on how IoT might change the management of supply chain:

- **Real-Time Location Monitoring:** The technology provides supply chain operations with a constant stream of actual data on item placement. If indeed the products are delivered incorrectly, you will be notified immediately and will be able to track the delivery until everything arrives at its location [17].

- **Monitoring the Storage Environment:** Supply chain managers may use particular sensors to monitor the transportation conditions of products, such as the temperature inside the vehicle, altitude, moisture, as well as other factors that may jeopardize service performance [18].

- **Determine the Travel and Delivery of the Merchandise:** The increasing number of options accessible to supply chain managers via IoT devices improves decision-making quality and increases delivery estimate accuracy. Real-time tracking allows a supply chain IoT firm to anticipate the ultimate delivery date as well as identify and eliminate hazards before they arise.

16.2.1.1 Enhance Emergency Preparedness

Smart IoT systems significantly assist supply chain managers in route planning, accounting for the fact unanticipated events on the road. The IoT collects all of the data required to develop a dynamic back up plan and avoid current delays [19].

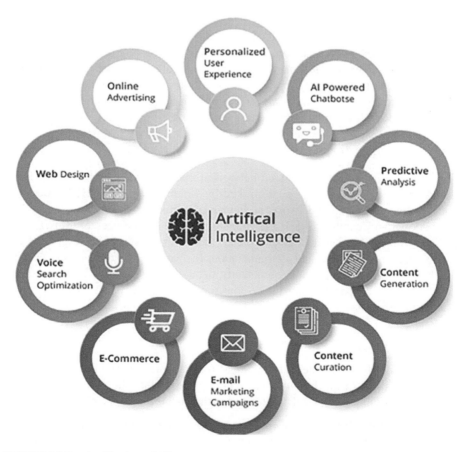

FIGURE 16.2 Application of AI.

16.2.2 Benefits of IoT for Management of Supply Chain

As the IoT grows in popularity, more companies are trying to embrace new technologies throughout the supply chain. Unless you're interested in how IoT design may boost your firm's efficiency and lower operational expenses, have a look at our list of IoT advantages for SCM [20].

- **Increased Speed:** The inclusion of path capabilities to an IoT SCM system enhances total supply chain speed substantially. The IoT shortens the continuous cycle, enabling for speedier judgment, less risk of delays, and improved productivity in identifying items inside the warehouse.
- **More adaptability:** When compared to on-premise systems, IoT solutions function faster and are more simply accessible by employing a cloud-based IoT platform. A linked IoT system sends all collected data in a manner suited for a specific circumstance, assisting workers as they navigate the distribution network [21].
- **Increased Precision:** IoT-based solutions offer managers with precise information on the lifetime of items, assisting merchants and supply chain managers in determining the appropriate manufacturing unit number to purchase. By aiding in the identification of products and directing drivers on their trip, IoT technology minimizes human errors [22].
- **Enhanced Efficiency:** It expands the capabilities of joined employee-oriented technologies. Smart glasses, for example, allow warehouse personnel to communicate with one another in real time, saving time when finishing a task. Furthermore, IoT captures performance data and improves knowledge of resource and labor management. The majority of companies continue to rely on outdated legacy systems to manage deliveries and monitor performance [22].

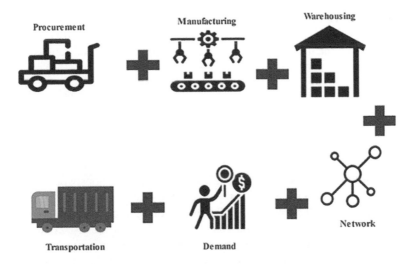

FIGURE 16.3 AI in supply chain management.

- **Lack of Skilled Team:** Adapting to the management of network platforms usually necessitates extensive training for warehouse staff and transportation drivers. It requires opportunity to address security rules and offer guidelines for using business platforms. Given the lack of IoT training, assembling a team capable of developing a solution that is tailored to the enterprise is challenging.
- **Problems with Increasing Data Storage:** The IoT has various advantages for logistics management, including big data pools. This benefit, however, comes with the requirement to buy enough servers to store and analyze all of the acquired data. To create appropriate reports using IoT-based insights, corporate executives must adopt information management policies and recruit data scientists and analysts [23].
- **Security Threats:** The construction of a secure IoT infrastructure is another task that supply chain managers must complete before migrating all processes to connected platforms. Vulnerabilities in computing and storage can result in threats and leaks, harming the company's performance and increasing the cost of failure.
- **Connectivity Problems:** A stable internet connection is critical for the IoT devices. There isn't always a continuous network because fleet drivers migrate from one region to another. Factors to consider while developing an IoT system for SCM is bandwidth. Connected systems sometimes necessitate a huge quantity of bandwidth and necessitate the use of a high-level server farm to work effectively.

16.3 ARTIFICIAL INTELLIGENCE

It is "the science behind the creation of intelligent machines, especially intelligent system software." It is similar to the equivalent job of using computers to understand human cognition, but AI is not limited to physically visible approaches. "Artificial intelligence (AI) is a vast subject of computer science concerned with the development of intelligent robots capable of doing tasks that would typically need human intelligence." "Artificial Intelligence has a wide range of applications, as listed below" [24].

16.3.1 ARTIFICIAL INTELLIGENCE (AI) IN SUPPLY CHAIN MANAGEMENT (SCM)

For quite some time now, AI appears to be the cure for nearly all professional and private concerns (from robotic labor to Alexa) and is creating a true start-up boom. Data collected from containers or manufacturing equipment via sensors should be used to predict issues in advance. Finally, AI should take over nearly the whole supply chain planning process and make decisions that were previously done by human professionals. To keep items moving properly, today's supply chain firms rely heavily on their wide network of suppliers and partners. To do so, they will require the appropriate technology to solve tactical and long-term concerns while also controlling the many risks associated with such complicated activities [25].

By incorporating AI into their supply chain, there appears to be a huge opportunity for innovation to reimagine how things are created, made, and delivered to customers.

Worker safety, predictive maintenance, finding process inefficiencies, and establishing intelligent supply chains that provide much bigger, higher quality items are all areas where AI can help. AI is the primary focus of SCM businesses due to its ability to automate, supplement, and improve customer experience and decision-making, as well as reconsider company strategy. In reality, the AI hype is encroaching on an area where the use of AI is already well established. Finally, it comes down to definition – in the public view, AI is now associated with methodologies like machine learning (ML) or even the subcategory deep learning. However, from a scientific standpoint, ML is simply one of several disciplines of AI. Existing SCM software effectively employs a variety of "methods" of AI, including ML, as well as entirely unrelated fields like operations research or fuzzy logic. Prediction and optimization are two common sub-applications in supply chain planning. Without delving into the specific benefits of ML for predicting, it can be stated that all approaches employed in this field process past data, such as sales and production levels, and offer a more or less trustworthy view for the future. This field of AI provides no answers for practical planning decisions, particularly the issue of what is the optimal action in a given scenario (optimization). Machine predicting only answers the question, "What may happen?" In other words, in terms of SCM, "What will the consumer demand for product XY look like in the following month?" or "How will the utilization of the other manufacturing machines evolve if machine A fails for 5 days owing to maintenance work?" [3].

At the end of the day, the SCM is still faced with a plethora of alternatives and must make the ultimate call. Knowing about future consumer demand opportunities does not always result in a single operational action. The forecast's output is simply one aspect in determining subsequent decisions, albeit a significant one. Here's an "extreme" illustration of the judgment difficulty in internal plant logistics: there are 3.6 million possible order sequences with 10 transportation orders (e.g., transporting a pallet from a materials warehouse to a manufacturing location) and one vehicle (e.g., a forklift truck). Modern supply chains are complex networks of companies, people, activities, data, and resources that are engaged in the movement of a product or service between supplier and customer. SCM solutions are generally manifested in software architecture systems that enable the flow of information across various departments inside and between business organizations. Leading SCM systems catalyze information sharing between organizational members and geographies, allowing judgment to get an enterprise-wide perspective of the information required in a fast, reliable, and systematic way. AI systems are using different kinds of cognitive computing, in addition to completely automated decision-making, to optimize their collective forces of artificial and natural intelligence. Intelligence in SCM, for example, provides greater supply chain automation through the use of computer programs that are employed either within (inside a single organization) and among supply chain stakeholders (e.g., customer-supplier chains).

16.3.2 How Can AI Be Utilized in the Supply Chain?

The supply chain is a large concept comprising several smaller activities and exchanges. These procedures offer a lot of potential for AI-powered SCM. Using ML in logistics management may help firms automate a variety of tedious operations, enabling them to focus on more critical business activities.

The present possibilities and uses of intelligent machines in the supply chain are as follows [26].

16.3.3   AI Possibilities in SCM

AI is a very important tool nowadays and it is so critical to transform Industry 4.0 to Industry 5.0. For any industry revolution, the major process is SCM. The next couple of sections describe the use of AI in this process.

16.3.3.1   Improve Human Workforces

Automation has the potential to significantly promote the development of leaner materials and production processes, which is critical for supply chains. Although most distribution and manufacturing jobs are now automated, incorporating technology gadgets powered by ML into such activities will significantly improve precision and agility. AI systems can also handle a wide range of warehouse concerns more quickly and precisely than people, simplify difficult operations, and enhance efficiency [3].

16.3.3.2   Supply and Demand Forecasting

Because of AI's capacity to manage vast volumes of data, it is useful for the important function of demand forecasting. AI anticipates economic demand in real time, accounting for the fact current sales, annual changes, and unusual alterations in consumer behavior. By forecasting inventory demands and adjusting throughout the network using pricing optimization algorithms, AI and system algorithms can keep supply chains learning. This data loop enables stock and supplier plans to be adjusted. Useful facts, scheduling, and distribution systems may now be altered to be proactive rather than reactive, thanks to AI technology.

Computer vision-based system
identifies faulty products

Identified defective products and parts
can be automatically sorted out

High-quality
Products

Defective products and parts

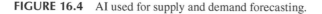

FIGURE 16.4   AI used for supply and demand forecasting.

16.3.3.3 Inventory Management (Turnover and Wastage)

Looking at simply the supply chain model and observing some of the numerous, diverse players who keep the logistics service network working is intriguing. A close investigation, however, will discover waste and underutilized processes. Consider food supply networks and the fact that around half of all food waste occurs when the voltage is set correctly. Important customers, manufacturers, distributors, and resellers across the supply chain retain extra inventory on hand than is technically necessary as a safety margin to ensure that all requests are completed without missing out on certain commodities. AI can assist in the effective design of supply networks. Supply and demand forecasting can aid in more precise planning, waste reduction, and cost reduction [27].

16.3.3.4 Quality Control and Smart Maintenance

Analyzing specific metrics using AI assists in forecasting, anticipating, and preventing quality issues, much as identifying subtle trends benefits in improving supply chain planning.

Companies may utilize image analysis as an example of how AI may be used to increase manufacturing accuracy. Visual inspections conducted directly upon that various manufacturing lines may identify trends that may go unnoticed in many activities. Sensor-based quality assurance systems, in addition to image processing, provide consistency and efficiency.

16.3.3.5 Shipping Efficiency

Groups working across the supply chain collect sensitive information (for example, requests, supplier status, production features, transit details, and so on), and this trend is expected to continue. Deep learning approaches enable machines to analyze real-time streams of data created by these modules in live time, resulting in faster modifications and improvements. "Artificial intelligence (AI) is a vast branch of computer science concerned with building intelligent robots capable of doing tasks that ordinarily require human intelligence" making the logistics process faster and more reliable. Logistics, such as ensuring that materials needed to complete a production arrive on time, is one of the most challenging difficulties that SCM faces.

Traditional stock levels are expedited by Intelligence solutions, which reduce operational bottlenecks along the value chain while requiring the least amount of work to fulfill delivery targets.

16.3.4 AI's Advantages in the Supply Chain Context

Implementing AI intelligently may have a number of practical consequences in domains such as:

16.3.4.1 Making Informed Decisions

By giving operational insights and comments on patterns and outliers, you may assist your organization in making better decisions and creating a leaner supply chain. You may also assist your staff in developing creative ideas and implementing data-driven choices.

16.3.4.2 Increased Efficiency

By leveraging AI and cognitive services, you may save money while automating your workers' tedious, repetitive duties. You can even discover issues before they occur. Identify bottlenecks and use automation technologies to speed up logistical processes.

16.3.4.3 Competitive Advantage

Make use of big data analysis to recover quickly while staying one competitive and successful: explore new opportunities and business practices, and improve logistics and operational.

16.3.4.4 Scaling Organization

You may enhance corporate development and scale your company by using AI to automate processes. AI and ML applications support the growth of global markets.

16.3.4.5 Customer Satisfaction

Improve client satisfaction by expediting shipping and having your items available within 24 hours. Make the entire process transparent to your consumers, with all status information available at all times, by facilitating human-computer contact via chatbots and speech recognition.

The Supply Chain's Artificial Intelligence Challenges AI adoption in the manufacturing process, as in other industries, is fraught with difficulties. To develop the IT infrastructure and get the data house in order, major investments, organizational changes, and migration from prior systems are required. Some solutions may need a large upfront investment (in terms of money, time, and operations) from a single organization and a large number of supply chain partners. Businesses must select the proper concerns and spend in creating and managing AI in order to secure its success. "Artificial intelligence (AI) is a vast subject of computer science concerned with the development of intelligent robots capable of doing tasks that ordinarily require human intelligence" [28]. The following are the most common challenges for adopting AI-based SCM solutions:

- Incorrect problem
- Inadequate data (or lack thereof)
- Legacy infrastructure

16.4 ORGANIZATIONAL TRANSFORMATIONS DISASTER MANAGEMENT

A catastrophe is a calamity that hits without warning and resulting in the casualties. Natural catastrophes and man-made disasters are the two sorts of disasters. It is an unexpected accident or natural disaster that causes significant damage or loss of life. Disaster management refers to actions done to ensure the safety and protection of people and property in the event of a natural or man-made disaster. When a tragedy strikes a society, foreign assistance in the form of aid is typically required to deal with the aftermath.

Preparedness: refers to actions taken before to a calamity. Preparedness strategies, emergency drills/trainings, and warning systems are some examples. Response: actions taken in the aftermath of a tragedy. For instance, public warning systems, emergency separations, and search and rescue.

Recovery: activities are those that take place after a calamity has occurred. For instance, temporary accommodation, claim processing and grants, long-term medical treatment, and counseling. Mitigation: actions that decrease the severity of a disaster's impact. Building codes and zoning, for example, as well as public education and vulnerability.

16.5   ROLE OF IOT IN DISASTER MANAGEMENT

While IoT technologies will not be able to prevent disasters, they will be extremely useful for preparations for disasters such as early warning systems and predictions before it happened. This technology will compensate for a lack of facilities in this way, putting underprivileged and developing countries at risk. Consider forest fire surveillance: sensors mounted on trees will collect data that indicates when a fire has started or a high risk exists, such as h temperature, humidity, moisture and different CO levels.

Different IoT applications, such as microwave sensors that will be used to analyze earth movements before and during earthquakes, or infrared sensors that will detect and monitor floods and human activities, have been created for other sorts of catastrophes.

IoT advancements may aid not just in disaster preparation, but also in disaster resilience. In the case of a disaster, careful planning of IoT-enabled devices (often

FIGURE 16.5   Various phases of disaster management.

battery-powered and capable of operating and communicating wirelessly) may provide advantages in terms of information network resiliency [32]. As a result, while catastrophe resilience is not their major goal, the side effect of offering a viable alternative telecommunications network may be extremely useful in places where the largest builder is weak, fragile, or non-existent, as illustrated in the following example. Thought leaders in this field are aggressively developing strategies to capitalize on technology advancements, and their impact will only grow with time. By incorporating these technologies into their strategic charter, agencies will reach unparalleled levels of speed, responsiveness, quality, and legality. During the stages of preparation, response, and recovery, the IoT has the potential to decimate disaster management [29]. Here are a few instances of transformative applications:

16.5.1 PREVENT

In the face of disasters, IoT is usually a game changer because of the following [31]:

- The employment of real-time sensing element-based data regularly significantly accelerates observance.
- Telematics-enabled automobiles.
- Water level sensors.
- Sensors for detecting wildfires, Hurricanes, tremors, cloudbursts, and volcanoes that are all examples of natural disasters.
- Critical infrastructure Condition monitoring of disaster risk management facilities offers protection.
- Risk reduction through the use of mistreatment sensors for pollutants and chemicals, as well as in hot environments.
- A facultative early warning observation system.

16.5.2 PREPARATION

The IoT has the ability to shape preparedness activities.

- Sensing element innovation is being used to deal with actual stock as well as provide replenishment,
- Designing spares and automatic indent process,
- Plus, track, and trace,
- For notification of associate degree action supported, a sophisticated event mechanism was used,
- Flowing sensor devices data is captured and used to prepare prognostic resources.

16.5.3 RESPONSE

IoT will facilitate response design and actions by:

- Vehicle tracking and geographic information system (GIS) integration
- Using sens ors to monitor key personnel movement

- Misusing near-field communication (NFC) for spatial enclosure and criterion barricades
- Situational awareness and incident management via information from one set
- Variety of information handling, prognostic analysis, big data, and complex event management
- Social media monitoring and processing

16.5.4 RECOVER

IoT is frequently a good facilitator for recovery efforts and activities by:

- Using sensing element technology for beneficiary identification and verification
- Using excellent cards and RFIDs for relief disbursal
- Creating a virtual provision Determent that enables hub operators and others to: stable competitive traffic toward and at periodic from a hub in real time; and
- Promoting communication between all parties involved.

16.6 CONCLUSION

In this chapter, we have discussed about the role of AI and IoT in supply chain man-agreement and disaster management. Also, we have discussed about the importance, benefits of these technologies. What are the challenges by usage of these technologies and how it is important to use them in SCM and disaster man-agreement. The AI platform improves results by evaluating and analyzing all available data and best practices in order to give insights and help decision-making. This not only helps to create a smart, responsive, and customer-centric supply chain, but it also improves safety, effectiveness, and transparency for all supply chain partners. AI can forecast supply chain demand, optimize routes, and regulate network management.

Furthermore, inventory management may be improved, resulting in increased profits and environment functions for the current infrastructure. It can further simplify and expedite the brokerage process, lowering costs and increasing error limits [30].

REFERENCES

[1] D. Anand, "Role of artificial intelligence in the field of medical and healthcare: A systematic review," *Annals of the Romanian Society for Cell Biology*, 3729–3737, 2021.
[2] M. H. R. Mehrizi, P. van Ooijen, and M. Homan, "Applications of artificial intelligence (AI) in diagnostic radiology: a technography study," *European Radiology*, 31, 4, 1805–1811, 2021.
[3] M. Pournader, H. Ghaderi, A. Hassanzadegan, and B. Fahimnia, "Artificial intelligence applications in supply chain management," *International Journal of Production Economics*, p. 108250, 2021.

[4] A. T. Rizvi, A. Haleem, S. Bahl, and M. Javaid, "Artificial intelligence (AI) and its applications in Indian manufacturing: A review," *Current Advances in Mechanical Engineering*, p. 825, 2021.

[5] Y. Riahi, T. Saikouk, A. Gunasekaran, and I. Badraoui, "Artificial intelligence applications in supply chain: A descriptive bibliometric analysis and future research directions." *Expert Systems with Applications*, 173, p. 114702, 2021.

[6] J. Bailey, "Inventive geniuses who changed the world: Fifty-three great British scientists and engineers and five centuries of innovation."

[7] L. Aldieri and C. P. Vinci, "Scalability and commercialization in support of sustainable development goals," *Industry, Innovation and Infrastructure*, 979–988, 2021.

[8] A. Wieland, "Dancing the supply chain: Toward transformative supply chain management," *Journal of Supply Chain Management*, 57, 1, 58–73, 2021.

[9] M. S. Sodhi and C. S. Tang, "Supply chain management for extreme conditions: Research opportunities," *Journal of Supply Chain Management*, 57, 1, 7–16, 2021.

[10] A. Zhang, J. X. Wang, M. Farooque, Y. Wang, and T.-M. Choi, "Multi-dimensional circular supply chain management: A comparative review of the state-of-the-art practices and research," *Transportation Research Part E: Logistics and Transportation Review*, 155, 102509, 2021.

[11] P. Goyal, A. K. Sahoo, T. K. Sharma, and P. K. Singh, "Internet of things: Applications, security and privacy: A survey," *Materials Today: Proceedings*, 34, 752–759, 2021.

[12] A. A. Laghari, K. Wu, R. A. Laghari, M. Ali, and A. A. Khan, "A review and state of art of internet of things (IoT)," *Archives of Computational Methods in Engineering*, 1–19, 2021.

[13] D. Anand and V. Khemchandani, "Unified and integrated authentication and key agreement scheme for e-governance system without verification table," *S͞adhana͞*, 44, 9, 1–14, 2019.

[14] D. Anand and V. Khemchandani, "Data security and privacy in 5g-enabled IoT," in *Blockchain for 5G-Enabled IoT*. Springer, 279–301, 2021.

[15] D. Anand, V. Khemchandani, M. Sabharawal, O. Cheikhrouhou, and O. Ben Fredj, "Lightweight technical implementation of single sign-on authentication and key agreement mechanism for multiserver architecture-based systems," *Security and Communication Networks*, 2021, 2021.

[16] D. Anand and V. Khemchandani, "Study of e-governance in India: A survey," *International Journal of Electronic Security and Digital Forensics*, 11, 2, 119–144, 2019.

[17] M. Ben-Daya, E. Hassini, and Z. Bahroun, "Internet of things and supply chain management: A literature review," *International Journal of Production Research*, 57, 15–16, 4719–4742, 2019.

[18] A. Sharma, J. Kaur, and I. Singh, "Internet of things (iot) in pharmaceutical manufacturing, warehousing, and supply chain management," *SN Computer Science*, 1, 4, 1–10, 2020.

[19] A. Aryal, Y. Liao, P. Nattuthurai, and B. Li, "The emerging big data analytics and IoT in supply chain management: A systematic review," *Supply Chain Management: An International Journal*, 2018.

[20] R. Arora, A. Haleem, and P. Arora, "Impact of IoT-enabled supply chain – A systematic literature review," in *Proceedings of International Conference in Mechanical and Energy Technology*. Springer, 2020, 513–518.

[21] N. Hemanth Kumar and S. Prabhudeva, "Layers based optimal privacy preservation of the on-premise data supported by the dual authentication and lightweight on fly encryption in cloud ecosystem," *Wireless Personal Communications*, 1–20, 2021.

[22] R. Dash, M. McMurtrey, C. Rebman, and U. K. Kar, "Application of artificial intelligence in automation of supply chain management," *Journal of Strategic Innovation and Sustainability*, 14, 3, 43–53, 2019.

[23] N. Singh, "Emerging technologies to support supply chain management," *Communications of the ACM*, 46, 9, 243–247, 2003.

[24] H. Min, "Artificial intelligence in supply chain management: Theory and applications," *International Journal of Logistics: Research and Applications*, 13, 1, 13–39, 2010.

[25] R. Toorajipour, V. Sohrabpour, A. Nazarpour, P. Oghazi, and M. Fischl, "Artificial intelligence in supply chain management: A systematic literature review," *Journal of Business Research*, 122, 502–517, 2021.

[26] B. Prasad, "Intelligent techniques for e-commerce." *Journal of Electronic Commerce Research*, 4, 2, 65–71, 2003.

[27] P. Lou, Z.-d. Zhou, Y.-P. Chen, and W. Ai, "Study on multi-agent-based agile supply chain management," *The International Journal of Advanced Manufacturing Technology*, 23, 3, 197–203, 2004.

[28] G. Baryannis, S. Validi, S. Dani, and G. Antoniou, "Supply chain risk management and artificial intelligence: state of the art and future research directions," *International Journal of Production Research*, 57, 7, 2179–2202, 2019.

[29] K. Sharma, D. Anand, M. Sabharwal, P. K. Tiwari, O. Cheikhrouhou, and T. Frikha, "A disaster management framework using internet of things-based interconnected devices," *Mathematical Problems in Engineering*, 2021, 2021.

[30] J. Xia, H. Jiang, and Q. Tang, "Introduction to artificial neural networks," in *Advanced Medical Statistics. World Scientific*, 1073–1090, 2003.

[31] Anand, A., S. P. Mishra, and S. Sahana. "Assistive devices and IoT in healthcare functions." *Deep Learning and IoT in Healthcare Systems: Paradigms and Applications*, 2021, 103.

[32] Yadav, L., S. Kumar, A. K. Sagar, and S. Sahana. "Architecture, applications and security for IOV: A survey." In *2018 International Conference on Advances in Computing, Communication Control and Networking (ICACCCN)*. IEEE, 383–390, 2018.

17 Cyborgs
A Coming Era

Anushka Purwar and Nidhi
University of Delhi

CONTENTS

17.1 INTRODUCTION

Cybernetic is one of the exciting areas of control systems and communication, including machines and living organisms. It studies the ability of humans, animals, and certain machines to adapt to or make modifications in response to environmental input. Cyborgs are created using the concept of cybernetics [1]. Philip Kennedy created the world's first cyborg in 1997. Philip Kennedy is known as the father of cyborg. There are around 300 people in the world who became Cyborgs by implementing themselves with a North sense on their chest that is attached with piercings. North sense is a device that starts vibrating whenever a person faces North direction. The computational capacity of biological systems are improved or replaced in cyborgs using artificial organs [2]. Are Artificial Intelligence (AI) and cybernetics the same? The answer is No. Researchers use AI that uses computer technology to create intelligent machines. Cybernetics practitioners utilize organizational models, feedback, objectives, and dialog to comprehend the capability and constraints of any system that can be technological, biological, or social; they see strong descriptions as the essential outcome [3].

DOI: 10.1201/9781003248750-17

A cyborg is a person whose physical processes are helped or controlled by technological equipment such as an oxygen tank, prosthetic heart valve, or insulin pump. Over time, the word has taken on a broader connotation, characterizing humanity's reliance on technology [1]. Although research on this subject is still in its early stages, cyborg intelligence offers many potential intriguing applications. At the intellectual level, cyborg intelligence poses a plethora of relevant and significant problems for AI research, and it has the potential to profoundly alter the AI landscape in multiple aspects [4]. This study aims to explore the cutting-edge technical application of AI or system intelligence via cyborg intelligence [5]. Brain-Computer Interface devices convert biological impulses collected from brain tissue into electrical signals used to control a computer interface. Once such control is achieved, it may be converted into a system capable of controlling a machine or a physical device, or even stimulating or activating biological tissues [6]. The examples of cyborg cockroaches and cyborg rats demonstrate cyborg implementation on insects and animals. Rat robots and Roboroach are common types of bio-robot [7]. The study on "cyborg enhancement technologies" focuses on the formation of new senses and enhancement of brain technologically. The advantage of such enhancement into brain includes direct information implantation, edition of memories, brain-to-brain wireless communication, and exploration and experimentation of sensory information in broad range [8]. The study closes with thoughts on the future of cyborgs and the meaning and consequences of becoming more cyborg and less human in an age of fast breakthroughs in the design and usage of computing technology.

17.2   BRAIN-COMPUTER INTERFACING: CYBERNETICS IN THE FIELD OF BIOTECHNOLOGY

Aging has a great impact on the human body and brain. It affects the molecules, cells, vasculature, gross morphology, and cognition. So, BCI (brain computing interface), sometimes called brain-machine interface (BMI), is now implanted into many patients' treatment. It provides a new communication channel for people suffering from cognitive and non-cognitive disease [5].

Researchers from Cornell University reported that they have developed the world's first non-invasive brain-to-brain interface known as BrainNet [9]. It combines both electroencephalography [4] and transcranial magnetic stimulation to collect brain impulses and transmit data to the brain.

BCI and BMI are somewhat the same but the technical difference between BCI and BMI is that of the output devices. In BMI, the translated signals are sent directly to the mechanical assistant device, but in BCI, signals are sent directly to the computer [6].

17.3   ROBOROACH: THE CYBORG COCKROACH

Are Roboroach new? No, they are not. For decades, scientists and researchers have been investigating bio-robotic platforms for insects. The Roboroach essentially incorporates a neuro-controller microcirculation created at UConn, which is a component of a tiny electronic backpack attached to the insect via wires connected to the insect's antennae lobes.

A cockroach moves to the left when a signal is sent to the right antenna. Similarly, if a charge is sent to the left antenna, the cockroach moves right. The information is sent to the operator by the microcircuit via the device's tiny Bluetooth antenna. An average cell phone can readily detect the signal; this is how a Roboroach works and we can control its movement from our own mobile devices.

17.4   CYBORG RAT

Researchers have created a very effective cyborg rat for loop resolution. The robot rat was built by implanting electrodes in its brain and strapping a portable micro-stimulator to its back. These devices allow a computer to communicate with the rat remotely, assisting it in finding the fastest course through a labyrinth, avoiding dead ends, and traversing loops. Rather than remotely controlling the rat, the computer provided ideas to assist the rat in making decisions [5].

Bio-robots have been used in a variety of animals, including cockroaches, rats, pigeons, goldfish, and others. Bio-robots outperform standard mechanical robots in many ways, including mobility, environmental adaptability, and energy usage, and they have enormous potential in rescuing and search operations [7]. Rat robots are a common type of bio-robot. An operator can use a computer to give control commands to the rat robot, causing it to follow a predefined path. Rat robots, however, are severely limited in practical applications due to their need for human instruction. When human supervision is unavailable in some rescue and search duties, such as under the remains of a fallen building, the rat robot is intended to be controlled automatically [7].

17.5   HOW ARE HUMANOID AND ROBOTS
DIFFERENT FROM CYBORGS

Humanoids and cyborgs are not the same things. Human intelligence is present in a cyborg, although it is largely based on robotics, whereas a humanoid incorporates AI in a human-like body [4]. Humanoid is also known as an android. Humanoids have detachable components, which means they may be updated with new parts or altered according to work at hand, but Cyborgs' parts cannot be removed because they are inside the human body. Like other machines and robots, humanoids can be reactivated, repaired, and reassembled [4], but Cyborgs cannot since they are human beings. Cyborgs have increased senses, powers, and talents, as well as computer-based minds, mental and physical capacities well beyond those of humans.

However, most people are unaware that cyborgs and robots exist, albeit not in the way that they are portrayed in films [10]. The existence of life is the primary distinction between a cyborg and a robot. A robot is essentially a highly advanced machine. It is frequently automated and requires relatively little human input. Cyborgs, however, are a biological creature combined with a machine. It doesn't have to be a person; it could be a dog, a bird, or any other living thing. A cyborg is distinguished from a robot by the presence of a living component. This indicates that a cyborg, but not a robot, is alive. A robot can only do what it was programmed to do, but cyborgs, particularly human cyborgs, have free choice over their actions [11]. Some officials are fearful that robots could one day become so sophisticated that they will be able to replace humans. An example of a robot is Pepper, a Japanese-built robot (semi-humanoid) capable of detecting and expressing a variety of emotions [4]. There are several good instances of robots. Among them are factory robots that do monotonous activities. Many robots outperform humans in these activities because they are faster and do not get exhausted [11]. Cyborgs live among us as well, but they may appear to be ordinary humans. Cyborgs are people who have robotic prosthetic parts. People who have pacemakers qualify as well, because their survival is partly dependent on the continued operation of the electronic device that regulates the rhythm of their heart [3].

17.6   SOME REAL-LIFE CYBORGS

17.6.1   HEARING COLOR: WORLD'S FAMOUS CYBORG

Neil Harbisson was born with the inability to see colors or color blindness. He sees everything in black and white [10]. He was unaware of his illness as a youngster until he was ten years old. Previously, his parents assumed that he just had trouble comprehending and describing color [12]. Once, his art professors let him sketch and paint in black, white, and gray when he was diagnosed.

He has an electronic antenna (Eyeborg antenna) attached to his lower back skull, which converts light frequencies into vibrations that his brain understands as sound, allowing him to hear hues such as infrared and ultraviolet (UV) [8]. He thinks that people are responsible for utilizing technology to alter themselves, and that the future is approaching in which technology will engulf humans in its surroundings [13]. He is a Catalonian artist and musician and co-founder of the Cyborg Foundation, where he helps become cyborgs [14]. Harbisson had to carry a 5-kg (11-pound) computer on his back that converted 6 fundamental color frequencies into 6 distinct tones. He remembered the name of the matching hue when he heard particular tones through a headset. He gradually increased the number of hues from 6–12 to 24–48. But there were certain issues; colors were converted into sine waves, and high-pitched electrical noises that gave him migraines. His body rejected constant sine waves and walked around all day with a hefty computer on his back. He kept tinkering with the Eyeborg, altering the size of the headphones and shrinking the computer [14]. Later, his chip was upgraded to incorporate a Bluetooth connection, allowing him to receive colors from other sources, such as phone calls. Infrared and UV sonic frequencies were added to his repertory. This gave him the remarkable ability to detect the presence of infrared sensors in a building or a terrible UV sunburn day. As he adapted to these new senses, his brain altered as well.

17.6.2   CAPTAIN CYBORG: DR. KEVIN WARWICK

Dr. Kevin Warwick is popularly known as "Captain Cyborg", as he is the first world's first cyborg. Since 1998, Warwick has been experimenting with different technological implants, including implanting a microchip in his arm, which allows him to remotely control lights, heaters, and computers [15]. As devoted as he is, Warwick implanted his wife with a device that allowed him to feel the identical feeling in his hand when someone grabbed hers. It's both jaw-dropping and oddly frightening.

17.7   ADVANTAGES AND DISADVANTAGES OF CYBORGS

There are several benefits of combining organic and mechanical components. The major benefit is to one's health. Many individuals are now classified as cyborgs as a result of surgical operations; many advancements in the medical profession have allowed people to be classified as cyborgs [10]. Many body parts can be replaced, including knees, wrists, elbows, veins, hip, heart valve, and arteries. In the coming era, as there will be more advancements in our technology, humans will be able to replace more body parts. Brain implants based on neuromorphic modeling are also available. Many studies have been carried out using cyborg technology, which has been applied to both humans and animals. Ratbot was an experiment in which a rat was fitted with a small electronic backpack and electrodes in its brain [11].

If there are pros of something, then there are cons as well. So, there are some drawbacks of combining organic and mechanical elements. Robots can use UV radiation, X-rays, infrared, and ultrasonic sensing to perceive things in ways that humans cannot. As a result, there is a greater reliance on cyber technology in general. Cyborgs are physically limited. Cyborgs are frequently incapable of healing bodily injuries and must instead be restored. For example, in the case of broken limbs and damaged armor plates, replacement is required, which can be costly and time-consuming [10]. Cyborgs have no trouble conceiving about the world around them in larger dimensions (multiple), but humans are more limited in this regard. In terms of memory and arithmetic processing, they outperform humans cognitively [11].

17.8   PROPOSED WORK AND METHODOLOGY

The survey was conducted on the awareness of cyborgs among students of age group 15–20 years, as this is the tender age group which can be the building block of future research and technology.

For the study purpose, the questionnaire was prepared, mentioned in Table 17.1, in which a total of 100 students participated.

The study was conducted in India, where the questionnaire was shared with students online. The participated candidates were from different colleges of the University of Delhi, India.

17.9   RESULT ANALYSIS

The result of the study for how many students are aware of Cyborgs has been studied and analyzed deeply. The result analyses of questions in Table 17.1 are mentioned below.

Hardly 29% of students in the age group of 15–20 years are aware of cyborgs, whereas 71% don't know about cyborgs, as depicted in Figure 17.1.

It is also concluded that about 54% of people think that humans are more powerful than Cyborgs, as shown in Figure 17.2, but in reality, Cyborgs are more powerful than humans, as they get themselves replaced with some organs made using AI.

TABLE 17.1
Questionnaire on Cyborgs

Sample size	100
Age group	15–20 years
Questions	1. Do you know about Cyborgs?
	2. Do you think that humans are more powerful than Cyborgs?
	3. Are you aware of Neuralink?
	4. Do you think Humanoids and Cyborgs are the same?
	5. Will Cyborgs replace humans in some years?

Do you know about Cyborgs? (Please be honest)
100 responses

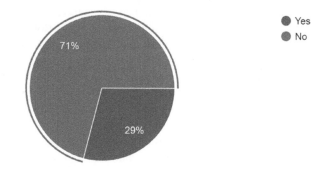

FIGURE 17.1   Awareness about cyborgs.

Do you think that humans are more powerful than Cyborgs?
100 responses

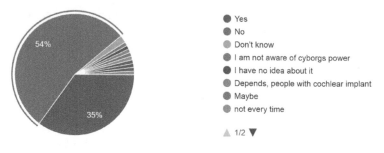

FIGURE 17.2   Human power comparison with cyborgs.

Are you aware about Neuralink?
94 responses

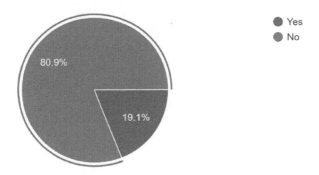

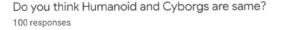

FIGURE 17.3   Awareness about Neuralink.

Do you think Humanoid and Cyborgs are same?
100 responses

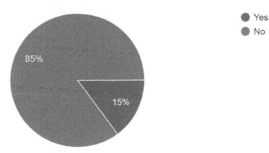

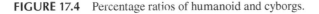

FIGURE 17.4   Percentage ratios of humanoid and cyborgs.

Figure 17.3 talks about the awareness of Neuralink in which it was found that 80.9% people are unaware of it compared to 19.1% of candidates who know about this recent technology.

Figure 17.4 states that many people know that Humanoids are different from Cyborgs but 15% of people think that they are the same.

The study also shows that about 59% of people believe that Cyborgs will not replace humans in some years, whereas 18% accept that Cyborgs will replace humans in future (Figure 17.1).

17.10   FUTURE OF CYBORGS

Cyborgs are slowly infiltrating our society. The most common depiction of a cyborg is that of a human-machine fusion found in science fiction films. A direct link between the nervous system and the Internet, such as the BMI, has been successfully tested in laboratory tests. However, the convergence of humans and machines is not limited to

Will Cyborgs replace humans in some years?
100 responses

- Yes
- No
- Maybe
- There are possibilities about that but...
- Don't know about cyborgs
- Don't know
- Since cyborgs are mostly humans wit...
- I am unaware about cyborgs

▲ 1/3 ▼

FIGURE 17.5   Cyborgs future prospects.

engineering technology. Biotechnology is one of the most important current technical areas. Individuals benefit from biotechnology through medical treatment, but we could be on the brink of a future in which people are made up entirely of biotechnology. Extending the natural limits of human lives or extending longevity is at the extreme end of improvement. For example, automobile technology allows us to travel at speeds that would be impossible to achieve in our natural state of leg movements, effectively extending and magnifying our natural state of being on foot [11]. Regenerative medicine can provide sick or dying people with better wellness or a longer lifespan through organ transplantation. As a result, biotechnology has aided people, such as regenerative medicine, and those who are thus made up of biotechnology are enhanced by this technology, which effectively extends their lives and instills the courage to battle death.

17.11   CONCLUSION

The study of communication and control systems involving living creatures and machines is known as cybernetics. The artificial parts used to build cyborgs use these to serve the primary function of an organ; can also add to, enhance, or replace the computational capacities of biological systems.

Cyborg implementation on insects is not new, many experiments were done on some insects, and these experiments are continuously going on to discover something new. Humanity is getting closer to merging with artificially intelligent robots thanks to a variety of cyborg technologies.

In this study, a comparative study on cyborgs has been analyzed. Different techniques have been studied and a survey was conducted on 100 people of age group of 15–20 years' students of various colleges of Delhi University, India. The analysis shows that only 29% of people are aware of cyborgs and 71% do not know that cyborgs even exist.

As many people are unaware of cyborgs and believe that robots, humanoids, and cyborgs are all the same, we have included some instances of the differences in this study. Real-life cyborgs are a wonderful illustration of how humans are less powerful than cyborgs. Furthermore, their organs can be replaced, recovered, or recharged as per the study suggested by prominent cyborgs mentioned in this study, such as Neil Harbisson, Dr. Kevin Warwick.

Cyborgs, bionic people, and robots with rising degrees of intelligence connect a series of intriguing subjects: cognitive, motor, and sensory prosthetics technology; biological and technical improvements to humans; and body hacking and brain-computer interfaces. According to the scientific research of Neuralink, which is a new technology. Indeed, the vision of the Human of Tomorrow is being discussed in transhumanistic projects, but good discussion necessitates an open approach to all perspectives to avoid narrow approaches.

In this chapter, we also discussed Cyborg Intelligence. Developing machines with human intelligence is one of AI research's most important and fruitful objectives. Finally, the benefits and drawbacks of cyborgs, a new impending era, were explored.

REFERENCES

[1] K. Sujatha, K. Sreekumar, I. Persis Urbana, B. Josiah Thomas, E. Anisha, and M. Karthika, "Cyborgs: Technical study," *International Journal of Pure and Applied Mathematics*, vol. 117, no. 22, pp. 221–224, 2017.

[2] E. Graham, "Cyborgs or goddesses? Becoming divine in a cyberfeminist age," *Information, Communication & Society*, vol. 2, no. 4, pp. 419–438, 1999.

[3] D. A. Mindell, "Cybernetics knowledge domains in engineering systems," *Computer Science*, vol. 2, p. 2011, 2000.

[4] W. Barfield, *Cyber-Humans: Our Future with Machines*. Springer International Publishing, Switzerland, 2015.

[5] D. Agushinta, F. Rindani, A. A. Kurniawan, F. Anggari, and R. Akbar, "Towards advanced development of cyborg intelligence," *Jurnal Ilmiah Informatika Komputer*, vol. 23, no. 3, pp. 201–211, 2020.

[6] C. Ogbujah and C. Ufomadu, "Ethical issues in cyborgization: The case of brain-computer interface," *Melintas*, vol. 26, no. 2, pp. 141–159, 2010.

[7] Y. Chen, H. Xu, W. Yang, C. Yang, and K. Xu, "Rat robot motion state identification based on a wearable inertial sensor," *Metrology and Measurement Systems*, vol. 28, no. 2, 2021.

[8] W. Barfield and A. Williams, "Cyborgs and enhancement technology," *Philosophies*, vol. 2, no. 1, p. 4, 2017.

[9] R. Kadyrov and I. Prokhorov, "Development automated biobot control system using AI framework," *Procedia Computer Science*, vol. 169, pp. 320–325, 2020.

[10] Dhandapani, J. P., "Cyborg technology: A quiet revolution," *Pondicherry Journal of Nursing*, vol. 12, no. 4, pp. 96–99, 2019.

[11] M. Sai Kumar, "Cyborgs – The future man kind," *International Journal of Scientific & Engineering Research*, vol. 5, pp. 414–420, 2014.

[12] G. Chakravarthi, *The Evolution of Synthetic Thought*, vol. 5 Takshashila Essay 2019-01, Takshashila Institution, Bangalore, 2019.

[13] R. H. Sagar, T. Ashraf, A. Sharma, K. Sai Raj Goud, S. Sahana, and A. K. Sagar. "Revolution of AI-enabled health care chat-bot system for patient assistance," In *Applications of Artificial Intelligence and Machine Learning*, pp. 229–249. Springer, Singapore, 2021.

[14] A. Almahameed, M. A. Oliva, and J. P. Borondo, "Cyborg acceptance in healthcare services: The use of cyborg as a surgeon," *Moving Technology Ethics at the Forefront of Society, Organizations and Governments*, Universidad de La Rioja, pp. 127–144, 2021.

[15] K. Warwick, "A study in cyborgs," *Ingenia, Journal of the Royal Academy of Engineering*, vol. 16, pp. 15–22, 2003.

Index